Paleopoetics

paleopoetics

THE EVOLUTION OF THE PRELITERATE IMAGINATION

Christopher Collins

Columbia University Press   New York

Columbia University Press
Publishers Since 1893
New York   Chichester, West Sussex
cup.columbia.edu
Copyright © 2013 Columbia University Press
Paperback edition, 2015
All rights reserved

Library of Congress Cataloging-in-Publication Data
Collins, Christopher.
  Paleopoetics : the evolution of the preliterate imagination / Christopher Collins.
      p. cm.
  Includes bibliographical references and index.
  ISBN 978-0-231-16092-6 (cloth)—ISBN 978-0-231-16093-3 (pbk.)—ISBN 978-0-231-53102-3 (ebook)
  1. Poetry—Psychological aspects. 2. Science and literature. 3. Cognition in literature. 4. Poetics—History—To 1500. 5. Discourse analysis, Literary. 6. Imagination in literature. I. Title.

  PN1083.S43C65 2013
  809.1'9356—dc23

2012029370

COVER PHOTO: Mariano Cecowski
COVER DESIGN: Martin N. Hinze

To the Ancestors,

all of them,

all the way back.

Contents

Preface ix
Some Notes on Dating and Nomenclature xv
Acknowledgments xvii

one The Idea of a Paleopoetics 1

  Relitigating Plato v. Poiêsis 2
  Big History 8
  Rhetoric, Poetics, and Hermeneutics 10
  The Presymbolic Mind 19
  Symbolic Play and the Verbal Artifact 22

two From Dualities to Dyads 28

  Duality 29
  Perception: The Parallel–Serial Dyad and Episodic Consciousness 34
  Action and the Anatomy of Multitasking 37
  Information Processing: The Parallel and Serial Modes 42
  The Dyadic Pattern 48

three Play and Instrumentality 57

  Donald's Four Stages of Consciousness 58
  Play in the Episodic Stage 62
  Instrumentality in the Mimetic Stage 68
  Imitative Play in the Mimetic Stage 72
  Preludes to Language 75

CONTENTS

four The World as We See It 82

    Vision and the Visual Imagination 83
    Parsing the Visible *Umwelt* 88
    Spatial Frames of Reference 93
    How *Homo* Became Sapient 96
    Complementarity—The Limits of Human Knowledge? 100

five Human Communication:
From Pre-Language to Protolanguage 106

    Why Language? 108
    From Pre-Language to Protolanguage and Beyond 114
    Gesture: Index and Icon 121
    Hearing Voices 129
    Protolanguage, the Long Transition 136

six Language: Its Prelinguistic Inheritance 141

    The Rhetorical Motive 142
    Language Play 146
    Verbal Visuality: The Simulation of Perception 153
    Verbal Visuality: The Simulation of Action 160
    The Rhetorical Imagination 168

seven The Poetics of the Verbal Artifact 175

    The Ritual and Poetic Genres 176
    Oral Performance Style 183
    Memory 189
    Enacting the Verbal Artifact 196
    Paralanguage, Protolanguage, and Oral Poetics 200

Epilogue: The Neopoetics of Writing 206

Notes 215
Bibliography 227
Index 247

Preface

The reader, it seems to me, is entitled to some explanation as to what in the world prompted the writing of a book like this, a book that claims to account for the prehistoric origins of literature. With that in mind, I will say a few words about the personal genesis of this unconventional project.

At some point in my mid-twenties it occurred to me that I would never be a great poet, probably not even a good poet. Yet writing a poem was still so magical an experience that I never wanted to give that up. And perhaps (I said to myself) I didn't have to give up that experience, that exultant high. Perhaps I simply needed to find a way to enter more intensely into the poetry of others. If I were to do so, however, I knew I had to lay aside much of what I had learned from my teachers, whose New Critical "thou-shalt-nots"—in particular, the "affective fallacy"—had ruled out as irrelevant the very thing I most valued, the reader's experience of the text.

I was only too happy to lay aside those dogmas, but I was already teaching college classes and finding myself engaged in those old routines of close reading and textual interpretation. How could I communicate my conviction that poems, short stories, and novels permit readers to participate in the writers' mind-altering process of creation and are therefore themselves a teachable art, if all I did was explicate phrases and clauses? At that time, though, I was confronting the experience of literature both from the perspective of the reader and of the writer, since, at New York University, I was teaching courses in British and American

PREFACE

literature *and* conducting poetry writing workshops. These two artistic activities, I hoped, might in some way be reconciled.

My first book, *The Act of Poetry* (Random House, 1970), tried to do just that. Here the common ground I staked out for reader and writer was memory: a writer arranges fragments of recollected events into a structure that readers use as though it represented their own remembered past. So, as readers, we don't read a piece of literature—*it reads us*—and this reading, like writing, is an *act*, not some written commentary on that event. Since one's memory is imprinted with sense experience, I could divide the book into chapters on visual, auditory, and motor/proprioceptive representations and thereby fit in all the basics: visual imagery (simile, metaphor, metonymy, personification, etc.), auditory effects (sound values, rhyme, etc.), and motor effects (rhythm, meter, line, stanza, etc.). My understanding of psychology then, I must confess, was rudimentary. As for my literary-theoretical position, it was anti–New Critical in all respects but one: I did believe that literary texts were single, unified structures, or at least became so in the mind of the reader. My heterodox views I'd absorbed partly from reading Blake, Whitman, and the Beats, partly from reading I. A. Richards, Gaston Bachelard, and Georges Poulet. When, later in the 1970s, I encountered Viktor Shklovsky, Louise Rosenblatt, Roman Ingarden, Roman Jakobson, Wolfgang Iser, and Michael Riffaterre, I discovered that what *The Act of Poetry* had been "doing" was Reader-Response theory.

Even before the Reader-Response movement began to lose steam at the onset of poststructuralism, it seemed to me to have made its separate peace with old-school literary interpretation. Instead of striving to identify the connections between stylistic features and mental events, it seemed now satisfied polling focus groups of readers and tabulating their interpretive choices. Some, like Norman Holland, then fed their results through a psychoanalytical filtering machine to analyze readers' preconceptions. Others, following the lead of Hans-Robert Jauss, concerned themselves with the variables that had determined the historical reception of given literary texts. Yet others, like Stanley Fish, attributed readers' interpretations to the influence of authoritative institutions. The questions they asked were often interesting, but they just weren't the questions I was interested in asking.

Then, in the late 1970s I became aware of a new kind of psychology. Having broken loose from the behaviorist establishment, it had devised experimental strategies that allowed researchers to draw inferences concerning what really goes on inside the "black box," the brain as the generator of consciousness. Consciousness, it seemed, had been a topic as taboo to behaviorism as a reader's thoughts and feelings had been to New Criticism. I learned that researchers, some of them, such as Robert Holt and George Sperling, working at NYU, were formulating what they called "Cognitive Psychology."

I soon found myself reading more papers on this new psychology than on literary studies—I remember once admitting to George Sperling that I'd become more familiar with his writings than with those of any of my English Department colleagues. At that time I was also drawn to the psychological writings of Endel Tulving, James J. Gibson, and Allan Paivio. Reading their papers, I found experiments that were clear and conclusions that seemed to promise answers to some of the oldest, most fundamental, and most perplexing questions in my field: What does fictive discourse do? How does verbally cued imagination work? What are the structures of memory and how do these relate to the structures of narrative? How are mood, feeling, and emotion generated in the reading of literature? The beauty of it all was that the scientists whose work seemed to be offering solutions to these literary conundrums were *not* literary theorists and, so, had blessedly no allegiance to this or that aesthetic school or "interpretive community."

In the early 1980s I participated in two NYU faculty colloquia. One, called the "Colloquium on Consciousness and the Brain," met twice monthly at the NYU Medical Center. It was chaired by Rodolfo Llinás and drew psychologists from the Washington Square campus, among whom I remember the young Tony Movshon. When after a year or so this colloquium adjourned for the last time, I convened one of my own and, joining with Paul Vitz of the Psychology Department, scheduled a series of talks and discussions on the relation of psychology to the arts. Named the "Psychoaesthetics Colloquium," it survived almost three years before it succumbed to the sort of malaise that too often besets such cross-disciplinary endeavors. Though interested and curious, participants from the humanities (e.g., literature, music, and the visual arts) and those from the various psychological subfields (e.g., cognitive, physiological, and educational psychology) were generally unready to assume another disciplinary perspective and learn to use its language. They were willing to showcase their findings but seemed reluctant to consider incorporating cross-disciplinary insights into their own ongoing research.

Though the ending of this discussion group was a disappointment, I continued to collect ideas from cognitive psychology, particularly ideas pertaining to visual imagery and applicable to the reading of literature.

PREFACE

These began to interact in my mind, as cross-disciplinary ideas tend to do when left to their own devices, and by the late 1980s I found I had brought to term a rather monstrous manuscript that I feared no editor would publish and, if published, no one else would read. I thereupon resolved to divide it surgically and produce from it two viable offspring. One, *Reading the Written Image*, traced the history of verbal images, analyzed the cultural biases against them, and proposed ways by which to read them. The last sentence of the book was a calculated nod toward my "next" book: "Any further study of the written image must squarely face the issue of mental imaging and must do so, it seems to me, within the disciplinary context of cognitive psychology." This companion opus, *Poetics of the Mind's Eye*, was wholly devoted to the application of cognitive models to the reading of literature, especially to the mental representation of verbal images as simulations of perception and memory. By pure coincidence, both were published simultaneously in 1991, one by Penn State University Press, the other by the University of Pennsylvania Press. The next year, when Reuven Tsur published *Toward a Theory of Cognitive Poetics*, the field I had been working in finally got a name. (Tsur, I must add, had published a 66-page paper in 1983, entitled "What is cognitive poetics?" but this question of his hadn't begun to provoke answers until the 1990s.)

In these two books I introduced a concept that will be central to this book, the concept that a verbal composition is an instrument, a verbal artifact. By "artifact," I don't mean what the New Critics meant by the "verbal icon," a quasi-spatial representation like a painting. A verbal artifact has an instrumental function: it is a skillfully made tool that we save and reuse to achieve some purpose. When it is not in use, it reverts to the status of an object and may be examined as an assembly of parts, but its meanings reveal themselves only in action, not in rest. For me the term "cognitive poetics" implies the study of the use of tools made of words.

Following the publication of my twin volumes in 1991, my interests drew me for a while from cognitive poetics to cognitive *rhetoric*. By "cognitive rhetoric" I mean the study of the persuasive resources inherent in the structure of language. As cognitive linguists and stylisticians identify them, these resources are the structural components out of which verbal artifacts are built, the process that is the special purview of cognitive *poetics*. (The working relationship of rhetoric and poetics will emerge as a major theme in the chapters that follow.) In my first venture into cognitive rhetoric, *Authority Figures* (1996), I examined two such rhetorical resources, the pronoun paradigm and metaphor, in the context of political authority within canonical texts from the *Iliad* to the Christian Book of Revelation. Continuing this line of research, *Homeland Mythology* (Penn State University Press, 2007) focused on the use of biblical narratives in the construction of American exceptionalism. In it I argued that metaphor, conceptual metaphor as George Lakoff defined it, has always been the basis of political authority and its founding myths, that this myth-making trope sleeps in language, waiting like a seed to germinate, or, an apter analogy, like a virus to replicate.

A fascination with the origins of human phenomena has itself an ancient origin. Every oral culture has had its origin myths and its tales of a majestic and unrecoverable age of heroes, which, when writing was introduced, became among its most revered documents. In later times this fascination has often been expressed in studies of mythic narratives as interpretable records of prehistoric behaviors, beliefs, and modes of consciousness. Perhaps because these tales were normally transmitted in heightened language and verse form, it has often been assumed that ancient humans were naturally "poetic." Because many of these early documents were preserved by priestly institutions and modified to honor local deities, it has also been assumed that these men and women were profoundly religious. In modern times this use of myths as guides to our deep past has led to a number of engrossing studies. One thinks of Johan Bachofen on matriarchy, James Frazer on sacred kingship, Carl Jung on mythic archetypes, Robert Graves on Celtic matriarchal poetics, and Julian Jaynes on the physiology of the Bronze Age brain.

My own current project, this "paleopoetics," takes a different tack. As an inquiry into human origins, specifically the cognitive/evolutionary origins of what we now know as "imaginative literature," it relies on information derived from much more recent sources, mainly from cognitive psychology and neuroscience. By living up to the name the U.S. Congress designated for it—the "Decade of the Brain"—the 1990s opened wide the gates of possibility for mine and for many another line of inquiry. Preceded by a quarter century of important discoveries, the split-brain findings among the most notable, the 1990s took full advantage of fresh technical breakthroughs, such as brain scanning devices, that accelerated advances in visual and auditory neuroscience, established the existence of mirror neurons that convert perception into simulated movement, proved that higher cognitive processes are performed by cells widely distributed throughout the brain, and disproved the old theory that brain cells, once they die, cannot be replaced. The brain, which we now know can regenerate its cells, continually reconnects them into new networks as new information is learned. Alongside these discoveries were those made in genetics (the Human Genome Project) and in computer technology (computer modeling and computational neuroscience). Through the 1990s and the early years of the new millennium, as the Internet provided researchers the means to share their ideas with colleagues, Francis Bacon's dream of the "New Atlantis" had finally become reality.

Though we may not yet agree on what selective pressures led our genus to evolve the particular cognitive skills it now possesses or ever know precisely when those skills emerged, we've come much closer to knowing *what* these adaptations must have been and the *sequence* of their appearance. Converging evidence from primatology, archaeology, paleoanthropology, genetics, neuroscience, linguistics, developmental psychology, and computer modeling has encouraged scholars to make claims that preceding generations would have dismissed as "just-so stories." This is especially the case for theories of the origins of language, a field that has flourished over the past twenty years and is, of course, the point of departure for this and any other theory of the origins of verbal artifacts.

Writers of prose fiction and poetry for their part have also been fascinated by these questions of origins and often saw their ancient counterparts as practitioners of an art that was simpler, stronger, and more genuine than that of later writers. The various literary genres of one's own age were like streams that one must trace upward to their fountainhead high in the mountains. The higher one climbed in this vision quest, the purer the water. In the early twentieth century, Ezra Pound was one writer who felt especially drawn to those ancient sources. In 1920, when he published a number of his literary essays under the title *Instigations*, he concluded with a posthumous essay by the sinologist, Ernest Fenollosa, "The Chinese Written Character as a Medium for Poetry." For Pound, the notion that a language of visual images could have preceded, or at least coexisted with, a spoken code seemed to confirm his deepest intuitions. That, and the prospect of traveling back in time to recover the originary power of poetry, he must have found most exciting in the essay. For there, among many other memorable insights, Fenollosa had written: "[P]oetry does consciously what the primitive races did unconsciously" and, therefore, the "chief purpose" of modern literary scholars and poets "lies in feeling back along the ancient lines of advance." The time has come when feeling back along those ancient lines has finally become a practical possibility.

Some Notes on Dating and Nomenclature

In recent decades paleoanthropologists have developed new means of dating hominid fossils. In matters of dating and nomenclature, I have throughout aimed for consensus and simplicity. In presenting evolutionary chronology, I have picked a median number between divergent estimates and/or rounded out the numbers: every date in the prehistoric record needs therefore to be preceded by an implicit "circa." Scholarly nomenclature has also adapted to new evidence and revised time lines. Based on genetic and molecular evidence, new taxonomic terms, e.g., *Homininae* and *Hominini*, have been introduced. While these may prove more correct, such terms are perversely similar. Since they characterize distinctions pertaining to *pre*human primate evolution, I have opted for the older, more familiar classificatory arrangement, i.e., *Primates*, or primate (order); *Hominidae*, or hominid (family); *Homo* (genus); *Homo sapiens* (species); and *Homo sapiens sapiens* (subspecies).

Acknowledgments

I want to thank the scholars who reviewed my manuscript: Peter Stockwell, Michael Corballis, Peter Schwenger, Claiborne Rice, Julie Kane, and Per Aage Brandt. While I may not have been able to incorporate all their suggestions in this final version, I have benefited immensely from having seen my project through their eyes. Their comments on its strengths and weaknesses guided me in the process of final revision. I am grateful also to Patrick Fitzgerald, Publisher for the Life Sciences, Columbia University Press, for his encouragement and apt insights; to assistant editor Bridget Flannery-McCoy, who skillfully shepherded my manuscript through the review process; to design assistant Katie Poe, who prepared the cover and the figures; to copyeditor Richard Camp, who carefully scanned my final draft; to production editor Pamela Nelson, who patiently inserted all my final revisions and prepared the entire project for the typesetters; and to my wife, Emily, who once again has done her very best to rein in my stylistic excesses.

Paleopoetics

one
The Idea of a Paleopoetics

How did it all begin? If some human activity especially fascinates us, we might become curious enough to ask that question. If that activity happens to be the reading of literature, our first impulse might be to think of the oldest preserved texts, such as the Chinese Book of Songs or the Vedic Hymns, portions of the Hebrew Bible or the Homeric epics. But we know these could not have been the earliest compositions. Thousands of years of preliterate chants, songs, and dramas must have preceded them. When we search for works of verbal art prior to these surviving texts, however, our eyes have nothing to peer into but what Prospero called "the dark backward and abysm of time." Archaeology can show us Neolithic textiles, Paleolithic figurines and cave paintings, 450-thousand-year-old wooden spears, and stone hand axes crafted some 2.5 million years ago (mya), but not one single prehistoric artifact made of words. True, the absence of evidence, as Carl Sagan liked to say, is not the evidence of absence (Sagan and Druyan, 1992:387). All the same, some small scrap of physical evidence that Pleistocene poets once roamed the earth would be reassuring.

The absence of material evidence is not the only challenge facing us. Even if we were to ask how the earliest *historic*—i.e., written—poetry came into being, we would still need to confront another vexing question: What do we mean by "poetry?" After all, we have to know what we are looking for. When *we* think of poetry, most of us think of lyric poems, printed texts in which rhythmical monologists express strong feelings as they struggle through problems to achieve moderately satisfying resolutions. Yet, for Aristotle, the man who gave us the word "poetics," poetry was not a thing or a set of cultural products but rather an activity, a "making" expressed in the verbal noun *poiêsis*. Moreover, it meant the making of narratives, dramas, and hymns, composed to be publicly performed—not the written lyric, our standard form of "poetry." And what of prose narrative fiction? What of the personal essay with its monologic voice searching for revelations and resolutions? What of prose poetry? What of unrhymed, unmetered free verse? What of folk songs, folk ballads, and children's rhymes? We recognize all these genres as somehow "poetic," but, as for "poetry," the object of our search seems to have been constantly changing over history into something else. One thing is certain: if the object of our search is *pre*historic poiêsis and the cognitive skills that must have made it possible, we must at the outset lay aside our literate conception of poetry as lines of words printed on white paper rectangles.

Relitigating Plato v. Poiêsis

We have many questions to ask as we undertake this search, but, before we do so, I must pose a metaquestion: What purpose would be served by answering those questions? That is, how would a theory of proto-poiêsis, a paleopoetics, affect the way we now understand and experience literature? Satisfying our curiosity about anything simply by weaving conjectures into a hypothesis, however artful the weave, is never ultimately satisfying. The only useful purpose of this or any study of origins is to shed new light on the objects under study and thereby encourage further research.

Despite their variety, what we recognize as imaginative compositions have some traits that seem regularly to recur. One of their traits is, for want of a better word, craziness, though perhaps they only *seem* to display that trait: the fact that verbal poiêsis, unlike the other arts, uses the medium upon which reason and logic are founded means that its moments of irrationality, when they do occur, seem all the more perverse. There is, on the other hand, a long tradition according to which poets, like prophets and shamans, are possessed by spiritual beings that speak through them. Plato, who banned all poets from his ideal republic because they told untruthful stories, explained in the *Ion* and the *Phaedrus* that they were also god-possessed madmen. According to the tradition he referred to, poetic utterances are the words of beings outside the human world that speak from within the bodies of humans, making poiêsis—verbal creation—both otherworldly *and* innerworldly.

Plato's Socrates, it should be recalled, left exiled Poiêsis the option to defend herself in court or to hire advocates to do so, a challenge that has prompted a series of writers—arguably Aristotle was the first—to reopen her case and appeal her sentence (*Republic*, book 10). (In English, Sir Philip Sidney and Percy Shelley also penned notable defenses.) If I, too, were to take up this long-standing challenge—and I suppose I am here doing so, after a fashion—I would begin by agreeing with Plato that poets do regularly deviate from rational discourse. But then I would argue that, when they do so, they do not go *out* of, nor do divine beings go *into*, their minds. Instead, poets go *into* their *own* minds and, doing so, guide us deeply into our own. To help establish that point, I would then proceed to call up a series of character witnesses, persons familiar with the defendant and able to share their insights with the court.

Consider how one extraordinary prose poet, Ralph Waldo Emerson, describes the effect words and thoughts could sometimes have on him:

> Underneath the inharmonious and trivial particulars [of our daily lives], is a musical perfection, the Ideal journeying always with us, the heaven without rent or seam. Do but observe the mode of our illumination. When I converse with a profound mind, or if at any time being alone I have good thoughts, I do not at once arrive at satisfactions, as when, being thirsty, I drink water, or go to the fire, being cold: no! but I am at first apprised of my vicinity to a new and excellent region of life. By persisting to read or to think, this region gives further sign of itself, as it were in flashes of light, in sudden discoveries of its profound beauty and repose, as if the clouds that covered it parted at intervals, and showed the approaching traveler the inland mountains, with the tranquil eternal meadows spread at their base, whereon flocks graze, and shepherds pipe and dance. But every insight from this realm of thought is felt as initial, and promises a sequel. I do not make it; I arrive there, and behold what was there already. I clap my hands in infantine joy and amazement, before the first opening to me of this august magnificence, old with the love and homage of innumerable ages, young with the life of life.... ("Experience," Emerson, 2001:207–8)

This experience seems to be above and outside him, a "heaven," and inside him, a mental space that opens into a vast world he has had no part in creating. He discovers it there where it has always been, a landscape immeasurably ancient, yet forever new, and, like a child he claps his hands with joy. A century later, Robert Duncan (1960) spoke of a similar visionary homeland in his poem "Often I Am Permitted to Return to a Meadow": it is "as if it were a scene made-up by the mind, / that is not mine, but is a made place, / that is mine. . . ."

After Emerson, I would introduce the sixteen-year-old Arthur Rimbaud, himself a poet of prose as well as verse, and have him describe his own mode of illumination: "It's wrong to say 'I think.' One should say: 'somebody thinks me.' Pardon the word play. I is someone else. . . . If brass wakes up as a trumpet, that's not its fault. To me that's obvious. I witness the unfolding of my thought: I look at it, I listen to it. I raise my bow to strike a note: the symphony begins to stir in the depths or comes leaping onto the stage." Though he may not have then read these latter passages, T. S. Eliot echoed them when he proposed his "impersonal theory of poetry" according to which the "progress of an artist is a continual self-sacrifice, a continual extinction of personality" until the poet becomes a medium, a crucible within which thoughts, feelings, and emotions are transmuted.[1]

In the Western tradition, the idea that each of us possesses, and is sometimes possessed by, some Inner Other derives (ironically) from Socrates' inner guardian spirit, his *daimonion*, later modulated by the Judeo-Christian body/soul dualism. This tradition was what Whitman used in order to articulate the curious relationship, sometimes discordant, sometimes erotic, between his inner and outer selves. For him his soul was the wise, deathless, visionary other, an entity that had already lived thousands of lives, while the body was the current, public self, the conscious identity that wore boots and a slouch hat, talked and sang, ate and drank and rode the Broadway omnibus. When poetry stirred within him, it was the soul that spoke, filling the body with a sudden influx of energy. The soul for Yeats was similarly ageless and energetic:

> An aged man is but a paltry thing,
> A tattered coat upon a stick, unless
> Soul clap its hands and sing, and louder sing
> For every tatter in its mortal dress.[2]

This language of spirit and inspiration is also linked to that tradition of spirit possession, which Plato spoke of as "divine madness," a tradition that includes the theology of prophecy with all its mysterious visitants, from the Muses of Hesiod, the *ruach* of Ezekiel, and Caedmon's dream-messenger to Lorca's *duende*, Graves's White Goddess, and the terrifying angels of Rilke.[3]

Reading, as well as writing, one can be overwhelmed by experiences that defy rational explanation. Consider Emily Dickinson's poetic touchstone: "If I read a book and it makes my whole body so cold no fire can ever warm me, I know that is poetry. If I feel physically as if the top of my head were taken off, I know that is poetry. These are the only ways I know it. Is there any other way?" For Ezra Pound the test was whether or not a set of words created in the reader what he called an "Image," a verbal pattern capable of presenting "an intellectual and emotional complex in an instant of time." Such a pattern, he continues, "is the presentation of such a 'complex' instantaneously which gives that sense of sudden liberation; that sense of freedom from time limits and space limits; that sense of sudden growth, which we experience in the presence of the greatest works of art." Pound's phrase "intellectual and emotional complex" reformulates the ancient claim that poiêsis uses rational discourse as a conduit for nonrational knowledge. This intuited knowledge, though prompted by language, would not itself consist of language. It would instead represent those prelinguistic processes associated with sensory input and motoric output, processes that, when they reach the level of thought, are often accompanied by emotion. T. S. Eliot (1919/1957) in his essay "The Metaphysical Poets" argued that intellect and emotion had, since the mid-seventeenth century, drifted so far apart that poetry had now assumed a sort of split personality, a "dissociation of sensibility" in which thought and feeling could no longer coexist. Poets needed to "find the verbal equivalent of states of mind and feeling." He concluded: "Those who object to the 'artificiality' of Milton or Dryden sometimes tell us to 'look into our hearts and write.' But that is not looking deep enough: Racine and Donne looked into a good deal more than the heart. One must look into the cerebral cortex, the nervous system, and the digestive tracts."[4]

If intellect and emotion can manage to merge "in an instant of time," the relatively slow serial processing of language must coexist with swift, "sudden," parallel-processed information associated with emotion. Whitman's defense of poetry, or the particular quotation from "Song of Myself" I would like to insert as his offering, is not a celebration of language or of the Platonic universes of discourse that philosophy can project. Instead, he creates a curious little unplatonic dialogue between himself and language. Language (speech) thinks that, just because it is able to represent anything and everything, it can override visuality (perception and imagination) and directly verbalize the inner self:

> My voice goes after what my eyes cannot reach,
> With the twirl of my tongue I encompass worlds, and volumes of
>   worlds.

But

> Speech is the twin of my vision, it is unequal to measure itself,
> It provokes me forever, it says sarcastically,
> *Walt you contain enough, why don't you let it out then?*
> Come now, I will not be tantalized, you conceive too much of
>   articulation.

Earlier in that text he had likened speech to the grass that grows on graves where, rooted in the breasts and mouths of the dead, it emerges as myriad tonguelike leaves. Now, elaborating that metaphor, he tells speech that it is like perennial grass whose rooted "live parts" survive the winter. His poems, formed of speech, may manifest themselves as leaves of grass, but their roots constitute an unspoken, unspeakable knowledge that corresponds to (tallies with) the ultimate meaning of things, a *rerum natura* that he equates with happiness. When this knowledge bursts open its buds in spring, what it utters is not language but some far more primordial sound:

> Do you not know O speech how the buds beneath you are folded?
> Waiting in gloom, protected by frost,
> The dirt receding before my prophetical screams;
> I underlying causes, to balance them at last,
> My knowledge my live parts, it keeping tally with the meaning of
>   things,
> Happiness, (which whoever hears me let him or her set out in search
>   of this day).[5]

If we agree that the primary purpose of language is to share with one another the knowledge we receive from nonverbal sources, i.e., internal sense data (what we feel and what we remember having felt) and external sense data (what we perceive in our environment), then language may be regarded as a sign system that mediates between these two fields of sensory reference. Whitman's metaphor suggests that this language art serves as a tally sheet to balance inner knowledge against the meaning of external things. Charles Olson also spoke to that point when he asserted that man as a "creature of nature (with certain instructions to carry out)" stands in direct relation to

> those other creations of nature which we may, with no derogation, call objects. For a man is himself an object, whatever he may take to be his advantages, the more likely to recognize himself as such the greater his advantages, particularly at that moment that he achieves an humilitas sufficient to make him of use. It comes to this: the use of a man, by himself and thus by others, lies in how he conceives his relation to nature, that force to which he owes his somewhat small existence. If he sprawl, he shall find little to sing but himself, and shall sing, nature has such paradoxical ways, by way of artificial forms outside of himself. But if he stays inside himself, if he is contained within his nature as he is participant in the larger force, he will be able to listen, and his hearing through himself will give him secrets objects share (Olson, 1950/1959).

That last phrase seems at first counterintuitive. Don't we humans share secrets with one another as *subjects* and do this through a language that is often self-referential and recursive ("What do you mean by 'thinks he doesn't know?'" "What I mean is that she thinks he doesn't know that her father really intends to . . . ," etc.)? But Olson is saying that the goal of poetic language is to tell the secrets that *objects* share with one another, a sharing by human and nonhuman objects that is essentially nonverbal. As Thomas McGrath wrote, "in the beginning was the *world*" (1982:287), and the word, when it did come, finally allowed humans to share with one another the overheard secrets of that world. This listening, which the Romantics meant when they spoke of "communing with Nature," science expresses as its faith in the intelligibility of the physical universe.

To a large extent, the dispute between philosophy and poetry that Plato instigated may be methodological. Philosophy, like science, is an open-ended activity, a conversation among opposing principles that manifests itself in "philosophizing." Poiêsis, on the other hand, manifests itself in compositions that may be sung, intoned, read, or acted out, verbal artifacts that, when performed from memory or silently perused, are objects that transform themselves into instruments by means of which

THE IDEA OF A PALEOPOETICS

their users extend their own powers of knowing. With that thought in mind, I might close this phase of my defense by introducing the testimony of William Carlos Williams on the vital knowledge—the "news"— that poiêsis has to communicate:

> Look at
> what passes for the new.
> You will not find it there but in
> despised poems.
> It is difficult
> to get the news from poems
> yet men die miserably every day
> for lack
> of what is found there.[6]

Big History

As a character witness, this man of science who in his student days dreamed of specializing in neurology, Dr. Williams provides a convenient segue to the body of my argument, which will be voiced by expert witnesses, men and women who claim no intimate knowledge of the defendant. Neither a hard nor soft scientist myself, what I will have to say about paleoanthropology, archaeology, psychology, neuroscience, and cognitive linguistics I have gathered as might an investigative reporter, convinced that there is a big, as yet untold, story out there that needs to be pieced together.

If this big story fits into any contemporary genre, it would be "Big History." As David Christian (1991) first defined it, this approach erases the traditional line dividing "history" from "prehistory," that is to say, accounts of the past derived from written documents as distinguished from those based on fossil and artifactual evidence. As he and others have argued, the discipline of history should mean the study of *all* past events. Big History should therefore begin with the Big Bang, currently estimated as having happened 13.7 billion years ago, and enlist input, first, from physicists, astronomers, chemists, and geologists, then from biologists, anthropologists, and psychologists, and finally, for events recorded over the past five thousand years, from traditionally trained historians. This division of labor corresponds generally to the hierarchy of sciences proposed by Auguste Comte in the mid-nineteenth century, a system recently reformulated by Gregg Henriques (2003) as the "Tree of Knowledge," according to which, for example, *physics* provides the basis upon which the principles of *chemistry* are built and chemistry provides the basis upon which the principles of *biology* are built.

Since I will confine my scope to the Quaternary period, ca. 2.5 mya to the present (i.e., the Pleistocene and Holocene epochs), my project might be termed "Not-Quite-So-Big History." Despite the relative modesty of my time scale, my project will seem to some no less vulnerable to the charge of reckless interdisciplinarity. To anticipate this I will quote Professor Christian:

> In tackling questions on these huge scales, the historian is bound to breach conventional discipline boundaries as well as conventional time scales. Can historians legitimately stray like this beyond their patch? Clearly, no single scholar can acquire an expert's knowledge in all the different disciplines that have a bearing on history at the very large scale. But this does not mean that the historian should abandon such questions. If a question requires some knowledge of biology or geology, then so be it.
>
> All that is required is a willingness to exploit the division of intellectual labor that exists in all our universities. Far from being unusual, this is normal procedure in any science; indeed it is normal procedure within and among the many sub-disciplines that make up history. Besides, such borrowing is more feasible today than it would have been even a decade ago; there exist now numerous fine works of popularization by specialists in many different academic disciplines, works that offer scholarly, up-to-date, and lucid summaries of the different fields. So there is no fundamental objection to the crossing of discipline boundaries; the difficulties are purely practical. (1991:226–27)

As for of my particular venture into Big History, its ultimate purpose is not simply to time-travel to earlier stages in the evolution of the human mind but rather to explore the depths you and I have within our minds *here and now*, those deep foundations within which certain strings of words have the power to resonate with astonishing results. Accordingly, my first premise is this: *the human brain is an embodiment of its own evolutionary narrative.* My second premise is that, broadly defined, *poetry is the brain's use of language to recover knowledge that is at once deeply past and deeply present.*

THE IDEA OF A PALEOPOETICS

The broad chronological model I will follow is that provided by Merlin Donald. In his *Origins of the Modern Mind* (1991) and *A Mind So Rare* (2001), he set forth a series of four stages that represent the evolution of hominid cognition from prehuman Australopithecine to contemporary human. I will review them in more detail in chapter 3, but at this point a brief outline is in order: (1) The *Episodic Stage*. The social circumstances of prehuman primates favored those individuals possessing a capacity to experience social encounters as complex, meaningful, ongoing episodes. This need, he suggests, put selective pressures on short-term working memory and rewarded those able to bind together longer and longer intervals, i.e., episodes, of experience. (2) The *Mimetic Stage*. With the advent of stone technology (ca. 2.5 mya), early humans demonstrated a degree of imitative aptitude that decisively separated them from their primate cousins. As the mimetic stage advanced, more and more cultural information was transmitted across generations, including firemaking and improved hunting techniques. (3) The *Mythic Stage*. While their immediate ancestors certainly had communicative skills, it was one small branch, *Homo sapiens sapiens,* that seems to have been the first to develop a full language (lexicon and syntax). The ability to communicate using a flexible, combinatorial system of spoken sounds continued a trend toward shared, culturally preserved knowledge. Finally, (4) The *Theoretic Stage*. This marked the full externalization of language in the form of written documents, the preservation of information without mnemonic structure, the dissemination of multiple copies, and the critical comparison of texts. It also marked a transition from biological to bio*cultural* evolution. Most evolutionary changes are cumulative, so, as Donald stresses, modern, literate humans are fully endowed with (1) episodic consciousness, the ability to monitor and assimilate up to an hour or more of data input; (2) mimetic capacity, the ability to observe and replicate the behavior of others; and (3) linguistic skill, the ability to use speech to share information, negotiate disputes, narrate past events, and plan future actions.

Rhetoric, Poetics, and Hermeneutics

This, then, accounts for the "paleo-" part of my title. Now, in order to clarify the "-poetics" part, I should start by offering a definition of poetics. Most broadly defined, poetics is the study of the principles and techniques of making things (from the Greek verb *poiein,* "to make"). Though it has sometimes been applied to the other arts, it usually refers to the verbal arts. In following this tradition, I will examine the principles that govern the making of a wide range of verbal artifacts, including folk tales, ballads, proverbs, rituals, epics, dramas, and novels, as well as those verse forms generally classified as "poetry." But in addition to the principles of making, I will also consider the poetics of *remaking*, i.e., the performance of such artifacts, either publicly before an audience or privately through the reading of written words.

But before I can explore its evolutionary implications, I need to distinguish poetics from two related verbal disciplines: rhetoric and hermeneutics. The problem is that these three share some of one another's properties, so, before I differentiate them, I must first understand how they are connected. To arrive at that understanding, I will borrow that hierarchical scheme familiar to the natural sciences. Accordingly, I will propose linguistics as the immediate basis of *rhetoric*, rhetoric as the basis of *poetics*, and poetics as the basis of *hermeneutics*: just as the various sciences deal with different emergent orders of material complexity, the latter three verbal disciplines deal with different emergent orders of symbolic complexity (M. Turner, 1991).

Language was the representational system our highly social ancestors devised to plan and execute cooperative action, to arbitrate disputes, and, from the point of view of the individual, to better manipulate the behavior of others to one's own advantage. These social purposes also entailed, of course, the sharing of object information, for knowledge about others (gossip) and knowledge about the environment (food sources and predators) would have increased one's value in the eyes of the community. Communicating social and object information in ways that convinced others of one's knowledge and trustworthiness became the valuable skill we understand as *rhetoric*.

Rhetoric, we should keep in mind, serves purposes that predate language and, in fact, predate the emergence of our human genus, purposes that include territorial dominance, sexual selection, alliance building, and all those other social negotiations practiced by our primate ancestors. This rhetoric would have been one of postural, gestural, and vocal signals.

When we view rhetoric as a means of achieving certain *human* goals, we classify it as a kind of tool. But this analogy needs a further distinction. Rhetoric is not *one* tool, nor is it a toolbox where a multitude of crafted implements are kept. It is rather a landscape strewn all over with findable tools. Since persuasive speech is most urgently required to cope with unforeseen circumstances, no one can know in advance what tool might be called for. The proper words must be found then and there and put together swiftly and effectively. If we think of language as a landscape, invention becomes the process of finding usable phrases, an activity that in a hunter-gatherer society would be analogous to finding particular stones, sticks, vines, or leaves that could be employed as ad hoc instruments to help crush, grind, poke, bind, wrap, or otherwise modify other things. In classical rhetoric this skill came to be termed "invention" (Latin, *inventio*; Greek *heurêsis*), the discovery of the verbal means to persuade an audience concerning a particular issue. The rhetorical devices we find do indeed serve us as found tools, and, once used, they are discarded—but not lost, because, if we have learned the art of rhetoric, we know the customary places, the *topoi*, or *loci*, in our mental landscape of language where we can go to retrieve them.

The devices that the rhetor, like some resourceful hunter-gatherer, nimbly retrieves are effective not only because they reverberate in language but also because they tap into our brain's prelinguistic strata. For example, narrative, including anecdote and exemplum, appeals to episodic memory, the principal means we have to organize our autobiographical past. Metaphor, metonymy, personification, and apostrophe are stylistic resources that evoke the powers of mental imaging and dream. Forceful delivery (*actio*) uses gesture and other paralinguistic features to excite incipient motor reactions in audience members, which in turn intensify their emotions.

Grounded in the art of rhetoric, poetics explores the next level of language-mediated complexity, that of closed, unitized, verbal systems, those entities that incorporate rhetorical devices while introducing their own special properties, e.g., meter, melody, dance, and mise-en-scène. To extend my instrumental analogy: poetics is the study of complex rhetorical artifacts, *made* tools that are *not* found through a process of improvisation and are *not* treated as disposable objects but are instead prized, saved, and reused.[7] In the first sentence of his *Poetics*, Aristotle implies that such works are made, like tools, in order to modify other things when he promises to consider each kind of poem in terms of its particular *dunamis*, i.e., its inherent property as an object to pass from a state of potency (*dunamis*) to a state of action (*energeia*). Poetics, as he went on to demonstrate in the body of his treatise, is a study of the ways in which verbal artists make structures of spoken words that, when activated by performers, can produce certain effects on an audience. When, therefore, he discussed *Oedipus Rex* or the *Iliad*, he did so in order to ask how such powerful works are *made* and what they *do*, not what they *mean*.

The fault line that would eventually separate poetics from hermeneutics appears as early as the fourth century B.C.E. Plato, in exiling poetry from his ideal state, does so on hermeneutic grounds: what poets mean does not correspond with verifiable facts.

Hermeneutics as a scholarly discipline, however, was slow to emerge. Aristotle, who seemed ready to write about virtually everything, never wrote a treatise on what we would recognize as hermeneutics. Though Plato critiqued Homer and his place in Greek pedagogy in the *Republic*, he never ventured into textual interpretation. Greeks and Romans quoted lines and sentences from literary works but were little given to commenting on those works. Granted, there were the periodic attempts to allegorize Homer, and several Late Latin commentaries on Vergil's poems were published, but classical readers seemed confident that they understood their writers without tutorial manuals.[8] It was not until culturally alien texts appeared, believed to be authored by an extraterrestrial god, that hermeneutics came into its own as a scholarly enterprise. Since then, the task of crafting a coherent theology out of a collection of disparate religious texts has taxed the hermeneutical ingenuity of two millennia of Christian scholars. Along the way, biblical exegesis, a specialized form of hermeneutics, has also provided humanist scholars with a repertoire of interpretive methods that they proceeded to apply, first to the Greek and Latin canon, then to vernacular European works, and eventually to selected works of world literature.

It was writing, *scriptura* in the generic sense of the word, that indirectly generated secular hermeneutics as a discipline. By the seventeenth century, writing and the book industry had made texts so numerous and so culturally diverse that, without expert interpretation, much of their meaning was inaccessible. Now, even within a common culture, a book written forty years ago may seem to young adults to be alien in its references, moral values, and affective tone. The artifacts that constitute the vast corpus of "literature" may still be tools, but most of them are not immediately usable. Now, when one of these tools arrives and is taken out of its box, it cannot be made to work until its pieces are put together and its operating procedures learned. And it has been the hermeneut's job, often in the format of a classroom lecture, to draft the how-to-assemble charts and write the user's manual.

As poetics emerged from rhetoric as the making, or *poiêsis*, of rhetorical artifacts, hermeneutics emerged from poetics to cope with the bewildering multiplication of artifacts that writing and eventually print culture produced. As a variety of technical writing, hermeneutics has now for well over a century been the principal discipline taught in every graduate and undergraduate department of literature. Such departments have simply assumed as their special mission the establishing of meaning and have pursued this one goal through a mixture of historical studies, textual analysis, and critical evaluation. Even those who have argued for the ambiguity or indeterminacy of meaning have addressed meaning as their central issue. How things made of words actually *do* what they do has seemed to most literary scholars an insufficiently important question to ask. Consequently, when it is not a synonym for versification, "poetics" has come to mean "literary theory," which has come to mean "critical theory," which, when professionally practiced, amounts to the interpretation of texts.

This activity is an appropriation of the text, as Paul Ricoeur (1991) called it. But one person's appropriation may be another person's misappropriation. Susan Sontag (1966/2001:7) wrote: "[I]nterpretation is the revenge of the intellect upon art. Even more. It is the revenge of the intellect upon the world. To interpret is to impoverish, to deplete the world—in order to set up a shadow world of 'meanings.' It is to turn *the* world into *this* world. ('This world'! As if there were any other.)"

Part of the disagreement may lie folded in the meanings of that word "interpretation." As I proposed (Collins 1991a:x–xxi; Collins 1991b:101–29), the verb "interpret" has two quite different meanings: (1) to perform a composition, as in the phrases "to interpret a dramatic role" or "interpret a piece of music," and (2) to convert an obscure message into a more understandable form, as one does by translating from one language into another. This is closely related to the duality of the gerund "reading," as (a) the act of reading and (b) the analysis of a text, as in "a reading."[9] That the very real differences between these two usages is usually ignored testifies to the depth of this problem. Interpretation #1 is an artistic enactment by which a scripted object passes from potency to act and, as such, lies in the purview of a *poetics*, whereas interpretation #2 is an analytical paraphrase that substitutes one coded message for another. When Stanley Fish (1982:355) announced that "interpretation, like it or not, is the only game in town," he was acknowledging a practice long established in scholarly and academic institutions, from the early Marxist, Freudian, and Jungian schools to New Criticism and from structuralism, through to poststructuralism, New Historicism, post-Colonial, and gender studies. While I would not deny that it serves a valuable pedagogical purpose,

I do not accept that hermeneutical interpretation should be—even if it still is—the only game in town.

My own attention, as I recount in my preface, has continued to be focused on the verbal work of art, not as an object of hermeneutical analysis, but rather as an instrument of cognitive action. The purpose of poetics, as I see it, is to study how that instrument is made and how the mind employs it, whether the verbal artifact is mediated by performers or by a written text. This experience, after all, is causally and, therefore, logically prior to literary interpretation, which can never be more insightful than the action of reading that precedes it. We should expect no less of literary interpretation than of travel writing, a genre that presupposes a real trip to, and real perceptions of, some real place. Being prior to hermeneutics, poetics cannot use hermeneutics as its disciplinary foundation, much less use hermeneutical practice to justify its own existence. It must build instead upon disciplines that are situated prior to itself, first rhetoric, then, continuing backward, linguistics, cognitive science, psychology, semiotics, and evolutionary biology.

Directly prior to linguistics in this order of disciplines, cognitive science emerged from psychology in the mid-1960s. Recruiting this new cognitive model, researchers first established the fact of mental images and their relation to visual perception, then proposed and tested hypotheses in neuroscience that have led to other far-ranging discoveries, all the while expanding their toolbox with the addition of computer modeling and brain imaging technologies. Thanks to its paradigm-changing revelations, cognitive science has suggested to a number of scientific disciplines, e.g., linguistics, anthropology, archaeology, and biology, certain new directions of research.

Its particular effect upon literary studies has, however, proved somewhat problematical. This became especially apparent when, in spring of 2002, *Poetics Today* published an entire issue devoted to "cognitive poetics," *Literature and the Cognitive Revolution*. Edited by Alan Richardson and Francis F. Steen, this issue presented contributions by Mark Turner, Paul Hernadi, Ellen Spolsky, Reuven Tsur, Lisa Zunshine, the two editors, and Tony Jackson, who concluded the issue with a commentary on the preceding articles.

Jackson's critique I found revealing for several reasons. For one thing, he asserted that the only justification for cognitive poetics was the degree to which it could improve the current practice of literary interpretation. The proof of any cognitive pudding was in the eating, which in his mind was the interpretation of an actual literary text, and, on that account, the results served up in *Poetics Today* he found generally unpalatable. What was so revolutionary about this "cognitive turn" in literary theory, if the writers who tried to demonstrate the usefulness of cognitive science to the explication of novels and poems could have reached the same interpretive outcomes had they used standard historicist methods? I confess I had to agree with Jackson on this last point. Reading "Literature and the Cognitive Revolution," I had kept imagining salespersons on some late-night TV infomercial touting some new and powerful device—a state-of-the-art computer that calculates the monthly grocery budget for a family of four or a surgical laser that juliennes string beans.[10]

I did, though, disagree with Jackson's premise and with that of most of the contributors as well, the notion that cognitive poetics should provide current hermeneutics with a revitalizing transfusion of new insights. This new and powerful model of the mind, it seemed to me, should first produce new and powerful insights into the mind's engagement with its own made instruments. If cognitive poetics does eventually transform the way we practice literary interpretation, it will only do so after it has wholly transformed the way we experience the works of art themselves.

But old institutional habits die slowly, if at all. In the Anglo-American tradition, the hermeneutical imperative came to be linked with a need to cleanse the interpretation of all elements deemed nonessential. As the New Critics diagnosed the problem, the objective thing-in-itself, the "verbal icon," could not be identified with the reader's experience of it, since that experience inevitably involved such messy features as idiosyncratic mental imagery and emotions. Of course, each reader's personal contributions to the experience will be different and may be adscititious, but, if these are unavoidable factors in every actual reading of a text, what sense does it make to demand that they cease to exist? A poetics that ignores the significance of nonverbal cognition, preferring instead to borrow a strict information-processing model from cognitive psychology, simply reframes the New Critics' "affective fallacy" by appealing to "higher" cortical processes and computational algorithms.[11]

The reader's brain does indeed process information in the form of verbal representations, but, equally important, it responds to them with its own internally generated *non*verbal representations, which manifest themselves as images, motor simulations, empathy, and a rich spectrum of affects—moods, feelings, and emotions. Words and verbal constructions do indeed direct these responses, but these responses do not resonate from the parts of the brain that process language. If anyone could actually read a piece of literature solely on the verbal level, the experience would be like only watching a pianist's fingers as they intricately move across a keyboard and never hearing the music. In short, when nonverbal processes become mere epiphenomena, the "embodied mind" loses its embodiment and cognitive poetics loses its right to call itself a poetics.

Cognitive poetics is not, however, the only project to re-envision humane letters in scientific terms. Since the publication of Joseph Carroll's *Evolution and Literary Theory* (1995), the movement known as "Literary Darwinism" has won for itself considerable visibility. Like cognitive poetics, it grew in reaction to the anti-science excesses of poststructuralist theory and the postmodernist notion that human behavior could be adequately explained by an analysis of cultural, as opposed to natural, factors. As cognitive poetics drew many of its early insights from cognitive psychology and cognitive linguistics, Literary Darwinism initially built upon two other sets of scientific ideas—sociobiology, as proposed by E. O. Wilson (1975), and evolutionary psychology, as defined by John Tooby and Leda Cosmides (1992).

As Carroll sees it, literature makes sense only in the light of evolution. Literary Darwinism, therefore, provides the only rational foundation literary studies can ever have, a programmatic position that has attracted a number of other literary scholars, such as Brian Boyd, Denis Dutton, and John Gottschall. In 2005 Carroll described this program as follows:

> In this hypothesis, the primary adaptive function of art is to provide the mind with subjectively weighted models of reality in such a way as to help organize the complex human motivational system. Art does not simply provide examples of appropriate behavior or adaptive information. It provides an emotionally saturated simulation of experience. Producing and consuming these simulations enable people both to experience the emotions depicted and to stand back from them and gain a cognitively detached sense of the larger patterns of human life. (This balancing between emotional involvement and cognitive detachment is what is meant by "aesthetic distance.") By vicariously participating in the simulated life provided by these models, people improve their ability to understand and regulate their own behavior and to assess the behavior of other people. (2005:940)

This, paragraph sums up his movement and proposes a hypothesis that, if verified—to the extent that it is verifiable—would indeed provide the humanities with a firm foundation. The problem comes when Literary Darwinists try to move from the summary level to the particulars.

Insofar as Literary Darwinism derives its explanatory power from evolutionary psychology, it posits a set of innate behavioral adaptations, e.g., survival instinct, sexual desire, competitiveness, kinship values, etc. Perhaps because any attempt to base these general traits in dedicated, domain-specific brain modules would be to balance a literary hypothesis atop a psychological hypothesis, Carroll has come to dissociate himself from what he now calls "'orthodox' or 'narrow-school' EP [evolutionary psychology]" and speaks instead in broader terms of a system of "elemental motives" (1999:409).[12] He thus detaches Literary Darwinism from the "massive modularity" thesis that has made evolutionary psychology so vulnerable in the eyes of its critics. Focusing its attention on the plot, Literary Darwinism connects overt narrative themes to "elemental motives," identified as ancestral adaptations that evolved over a period of 1.5 million years. In short, since Pleistocene adaptations still govern the choices of fictional characters, Darwinism provides the only foolproof key to literary meaning.

Thematics is a quite valid field of inquiry. It is important, from time to time, to stop and ask ourselves: Why does this playwright or that novelist portray characters perplexed by this or that dilemma or driven by this or that passion? The presence of given themes in a literary text may indeed reflect adaptations that were once necessary for the survival of our hominid ancestors, but this is not a sufficient explanation for their survival in fictive scenarios or indeed for the survival of fictive scenarios themselves. As Jonathan Kramnick (2011) proposed in his recent critique of Literary Darwinism, in reading pieces of fiction there are more complex processes at work involving pretense, imagination, memory, emotion, and other "features of mind [evolutionarily] selected (if at all) for *other* purposes." It is not sufficient simply to restate a set of Pleistocene adaptations in a manner that is "relentlessly *thematic*" (2011:340, 344; author's emphases).

Kramnick's reference to pretense is especially apt in light of Carroll's curious claim that "fictionality is not a distinguishing characteristic of literature" (1995:107). If one accepted that claim, all questions of fictive play and defamiliarizing representations would have to be ruled out in advance (Miall, 2005:146). But declaring, in effect, "It's just that simple!" and ruling out troublesome questions only provokes other cage-rattling questions, such as: Are hermeneutic systems, like literary texts, also driven by "elemental motives?" If so, what "elemental motive" is responsible for Literary Darwinism? Was there a point in time that a passion to dominate Paleolithic palaver became selected as an inheritable trait?

Just that simple? No, not just *that* simple.

The Presymbolic Mind

This book represents an alternative way to bring poetics into alignment with human evolution. Rather than focus on narrative themes, I examine in detail various cognitive skills essential to verbal art and trace their gradual emergence over the long prehistory of our species. These are skills adapted to the presymbolic mind. That is, they evolved before humans communicated with one another using arbitrary signs—conventional symbols—and relied on visual images for both inner thought and the outward exchange of thoughts. Note: Throughout this book I have consistently used "symbol" and "symbolic" as semiotic terms referring to arbitrary signs, such as words (spoken and written), as distinct from indices (signs that refer to physically associated objects) and icons (signs that resemble what they refer to). I do *not* use "symbol" in the sense of an image that stands for a complex structure of hidden meanings, a usage common to religion, psychoanalysis, and literary studies.

Chapter 2, "From Dualities to Dyads," begins my exploration of cognitive skills preadaptive to language with a consideration of Dual-Process Theory. This field, which emerged in the mid-1990s, posits the coexistence in the brain of two distinct cognitive systems, one intuitive, the other deliberative, a bipartite arrangement reminiscent of the right-hemisphere/left-hemisphere duality. Since the intuitive system comprises cognitive features shared with nonhuman animals and the deliberative system is uniquely human, the evolutionary implications of Dual-Process Theory make it highly relevant to paleopoetics.

I note that this is a restrictively cognitive theory, by which I mean it focuses on information processing to the exclusion of that overarching duality, perception and action. In an effort to incorporate these larger functions, I propose that parallel and serial processes, which dual-process theorists list among their paired opposites, are not only essential to information processing but are also key modes of perception and action. In perception, for example, they collaborate to produce figure–ground discrimination; in action they collaborate to effect multitasking. The parallel and serial modes are therefore not contraries but rather correlatives, necessary complements that come together to constitute an integrated duality, or, as I call it, a "dyad." The distinctive pattern they present is one in which broad and diffuse awareness coexists with narrow and finely defined attention. Replicated in diverse functions, from bimanual coordination to episodic memory, this dyadic pattern provides a heuristic key to yet other functions, such as gestural communication, protolanguage, full language, and, ultimately, verbal artifacts. In proposing this hypothesis as a friendly amendment to the Dual-Process Theory, I suggest that the dyadic pattern was an adaptation selected to smooth the transition from the prehuman to the fully human brain, a transition not yet completed.

In chapter 3, "Play and Instrumentality," I examine these two elements as presymbolic functions preadaptive to language. As such, they are associated with Donald's episodic and mimetic stages, respectively. Observed especially among young mammals, play has been variously explained as a way to exercise hunting and fight-or-flight routines, to bond with peers, to establish status in a social hierarchy, or simply to discharge excess energy. Play undoubtedly does serve these particular purposes, but, taking my cue from Gregory Bateson's essay "A Theory of Play and Fantasy" (1972), I regard play more broadly as a central and formative factor in cognitive evolution. As Bateson observed, play requires participants to enclose certain actions inside a frame, a "play frame" within which an animal's actions (e.g., chasing, kicking, and biting) that normally signify X, now signify Y. In other words, play participants interpret indexical signs as merely icons (semblances) of indices, and so, in this case, aggressive behavior only *seems* aggressive. By combining icon and index, social play becomes an instance of dyadic patterning, one that also produces another effect: it detaches a natural sign from its normal significance and converts that sign into an intentional token usable in itself as a communicable thought.

Of course, signs that can float about detached from jointly perceived contexts may become means of deception. For that reason social play requires a high degree of trust. The sign systems that evolved within genus *Homo*, beginning perhaps some 2.5 mya, succeeded only because in-group cohesion was strong enough to minimize deception and maximize trust. Whenever a conventional system of signs, a lexicon of arbitrary symbols, emerged and, by about 100,000 years ago, was organized by a compositional syntax, it owed its existence to the play instinct, that old mammalian trick of "We *do this*, but we *mean that*." In linguistic terms, this amounts to, for example, "We say '*dog*,' but we mean '*that short furry animal that wags its tail and barks.'"* The "this-means-that" semiotic play principle from which every symbolic sign system flows is also the principle that first breathed life into the verbal artifact as an imitative performance. Storytellers, dramatic actors, and ritual performers initiated their work by stepping inside a play frame, a kind of sacred space in which ordinary words and actions came to assume extraordinary meanings. As Coleridge put it, the first principle of "poetic faith" is a "suspension of disbelief," a quasi-religious affirmation ("for the moment") of the truth of this performance.[13]

The second function I deal with in this chapter is *instrumentality*, by which I mean the use of an object as a tool to enhance human action. The magical potential of an entity to transform itself from an external thing to an extension of the user is the essence of the object–tool dyad. But here another distinction is evolutionarily crucial, the nondyadic distinction between *found* and *made* tools. When tasks only required clubs, reaching sticks, or nut-cracking stones, the materials for such tools might be readily found when needed, but when the tasks required cutting, grinding, shaving, binding, or sewing, proper tools would have to be crafted and then preserved for future use. The reusable instrument, so central to the notion of material culture, was the prototype from which other less palpable instruments were to be fashioned and preserved. These latter instruments we may call cultural behaviors, repeatable sequences of action that left no archaeological traces and can only be inferred from physical evidence, e.g., stone tools and fossilized bones. Among those behaviors I include tool making, tool use, dances, and songs. The latter, as sequences of symbolic sounds, I designate as *verbal artifacts*.

The last presymbolic function I discuss is *visuality*, a topic to which I devote the whole of chapter 4. I chose to entitle this chapter "The World as We See It" in order to suggest that so much of what we understand as our terrestrial environment is determined by the ways our visual anatomy construes it. From the surface of our eyes to the various deep pathways of the brain, visual data, as they pass along, are given structures that have come to characterize what Jakob von Uexküll (1921) would term our human *umwelt*. In this chapter I investigate the neural circuitry by which the brain processes visual arrays and show how this sense modality helps coordinate locomotion, prehension, and other essential actions. The discovery (1980–1995) of the two visual pathways, or streams, revealed how serial and parallel processes collaborate to produce an integrated visual field, including figure–ground distinctions. It further illustrated the capacity of the brain to simultaneously perceive (1) a sharply defined, narrowly focused figure and (2) a broadly scanned, relatively diffuse ground of other objects.

When I chose for my subtitle the phrase "the evolution of the preliterate imagination," I meant these words quite literally, especially that word "imagination." I did not mean the phrase as a high-sounding synonym for know-how or cleverness. I meant it as the simulation of sensory perception, especially visual perception. Unless we have a clear understanding of visual perception, both as a means of selecting and recognizing objects and as a means of interacting with them, our understanding of simulated vision, i.e., imagination, will be rudimentary. Since one of the principal achievements of language is the encoding of our visible umwelt as verbal images, I have reviewed here some of the findings of current neuroscience that seem to me to be most pertinent to the study of literature.

Symbolic Play and the Verbal Artifact

With chapter 4, I complete my preliminary presentation of the dyadic features of the presymbolic mind that I have inserted into Merlin Donald's chronology, claiming them as preadaptive to language and its artifacts. Having introduced the issues of instrumentality, play, and visuality I can now venture an anticipatory definition of the verbal artifact as an *instrument that the brain uses to play visual mental images*. If we apply the terms of this definition to the other so-called sister arts, we can see that, while all have play in common, important specific differences exist. In dance we may say that the body is the instrument, but we may not do so if we define an instrument as an external tool, a prosthesis. In nonverbal music, external instrumentation is customarily used, but the channel is auditory, not visual-imaginal. Finally, in painting and sculpture we use a visual object as an external instrument, but what results from that are visual perceptions, not mental images.

Verbal artifacture—poetry, broadly defined—has language as its instrumental medium. It is this communicative code of arbitrary symbols, its origins, and its preliterate uses that I investigate in the final three chapters of this book. Modern glottogony, the theory of language origins, was hastily conceived in the mid-nineteenth century in a liaison of linguistics and evolutionary biology, prematurely born, and, soon after, nearly smothered in its cradle by the *Société de Linguistique de Paris* and the Sanskrit philologist Friedrich Max Müller. The French society in

1866 announced it would no longer consider publishing articles that dealt with the origin of language. (And, it must be said, most of the work then published in this nascent field ranged from the implausible to the preposterous). Max Müller (1868) ridiculed any and all theories that derived language from nonlinguistic behaviors, e.g., the imitation of sounds to represent their sources (the "bow-wow" theory), emotive interjection (the "pooh-pooh" theory), and sounds used to coordinate collective action (the "yo-heave-ho" theory). All Darwinian speculation was wrong from the start because, as he averred, "language forms an unpassable barrier between man and beast" (Müller, 1889).

It took a century for an evolution-based glottogony to come of age and another half century for it to make substantial advances. Noam Chomsky (1968) introduced the idea that human language could best be explained as the result of a genetic mutation that occurred at some point early in the emergence of *Homo sapiens* as a distinct species. Ironically, though, Chomsky and those who followed his lead, e.g., Derek Bickerton (1992) and Steven Pinker (1994), still managed to keep Max Müller's unpassable barrier firmly in place by excluding any possibility of intermediate stages between nonlinguistic communication and fully grammatical language.

It was therefore left to other scholars to argue the case for some form of prelinguistic communication variously based on gesture and noncompositional vocalization. It is principally their ideas that I review in chapter 5, "Human Communication: From Pre-Language to Protolanguage." With this chapter I also return to the topic of mimesis and the evolutionary traits associated with Donald's mimetic stage. I begin this discussion with the much-debated question of whether social information or object information came first. Social information is communication that, anchored in a present circumstance, is intended to influence the behavior of others; object information uses names to identify objects and may refer to nonpresent circumstances—the past, the future, and the elsewhere. If social information came first, the link between primate communication and language remains intact, for all primates have gestural and vocal means of expressing their needs and fears and use this wordless rhetoric to influence the behavior of conspecifics. Conversely, if object information came first, the unpassable barrier remains in place, for the capacity to employ a vocabulary of referential names would have to have appeared suddenly and fully formed. The majority opinion now seems to favor social information as the preadaptive matrix from which object information gradually emerged. This means that gestures, online indexical signs such as pointing and iconic signs such as handshapes, would have preceded the use of arbitrary symbols for displaced referents.

A symbolic system, gestural perhaps at first but eventually vocal, would have permitted humans to refer to, and think about, absent things and events. Did a symbolic communicative code, a vocal lexicon without a syntax, precede the emergence of language? Was this "protolanguage" a system of gestures, of vocalizations, or a combination of both? Could humans have used it to communicate without some, albeit rudimentary, syntax? These questions have provoked a lively debate over the past two decades. The current estimates as to when full language came into being vary greatly, but, since all humans possess it, we generally assume it appeared prior to 60,000 B.P. (before the present) when our species began to diverge in its migration out of Africa.

After presenting a time line that incorporates a number of theoretical positions, I analyze gesture as a semiotic medium and note that, oddly enough, writers in this field have seldom taken advantage of the simple distinctions of index, icon, and symbol with two notable exceptions, Terrence Deacon (1997) and Jordan Zlatev (2008). The conclusions they arrive at are, however, strikingly opposed. Deacon uses the Peircean distinctions in order to reinforce Max Müller's barrier by claiming that in no way can symbolic signs be generated from icons and indices. Zlatev, on the other hand, sees the latter two sign functions as preadaptive to a symbolic code, such as language. I go on to support Zlatev's position by suggesting several ways by which selective pressures could indeed have driven the human communicative code from index and icon to symbol and then extended its medium from gesture to voice. With a second time line I illustrate how this transition could have occurred and propose that a gradually changing protolanguage could have been in use during a transitional period of over 500,000 years.

With the emergence of full, or true, language, we arrive at the onset of Donald's mythic stage. Chapter 6, titled "Language: Its Prelinguistic Inheritance," highlights those cognitive aspects of pre-language that were later selected for the more persuasive functions of rhetoric and for the shaping of verbal artifacts. One prelinguistic aspect is the centrality of a dominant individual, represented in the pronoun paradigm. As I proposed in *Authority Figures* (1996), this rhetorical structure represents the speaker (*I*) as the central visual/vocal source of information in respect to an audience (*You*) and excludes from this speech circle all third persons (*They*). In turn-taking conversation the roles of first and second persons shift, but in poetic performance, whatever its form, an unchanging speaker-to-hearer relationship tends to be maintained.

I next reintroduce two themes discussed earlier, play and visuality, and place them in the context of cognitive linguistics and its emergent subfield, cognitive rhetoric. The relation of symbol to its referent, unlike that of index and icon to their referents, is arbitrary. Having no sensory connection with its meaning, a word may justifiably be termed *nonsensical*. It is only the play instinct with its double-framing that can maintain a rule-governed connection of the symbolic signifier to its signified. The fact that infants during those crucial months of early language acquisition are also acquiring the principle of pretend-play suggests that the dyadic pattern constitutive of social play may be preadaptive to the dyadic pattern that links arbitrary sounds with intended referents. In exploring this linkage, I also review George Lakoff's theory of conceptual metaphor and argue that metaphor and metonymy are also rhetorical expressions of mammalian play behavior.

Having commented on the play aspects of lexical elements, I then turn to the visual aspects of syntax and, citing the work of Talmy Givón, Ronald Langacker, and Leonard Talmy, consider how language functions as a means of simulating visual perception of objects and visually guided action. Referring back to chapter 4, as well as to my book *Poetics of the Mind's Eye* (1991a), I align the traditional parts of speech with specific optical processes, e.g., fixations and saccades. I then return to the dual-pathway theory of vision and map it onto some of the models proposed by cognitive linguists. I conclude this chapter by proposing that the five classical canons of rhetoric, particularly invention and style, represent vision-based cognitive functions that also operate in spontaneous speech.

In my final chapter, "The Poetics of the Verbal Artifact," I suggest how the earliest kinds of poetry may have emerged as formularies used in various rituals pertaining to life events (birth, coming of age, marriage, death, etc.) and food production cycles (hunting, fishing, sowing, harvesting, etc.). These spoken portions would include origin myths, prayers, and incantations that might later become detached from these ritual actions.

The rhetorical features, inherent in language, would now be used in these carefully made instruments of words to focus a hearer's attention by holding and extending the duration of his or her short-term working memory. Verbal artifacts would also need to incorporate elements that facilitate long-term memory storage between performances. The major stylistic features associated with oral composition, e.g., repetition, formulas, parataxis, and extraordinary events, all exist to serve the needs of these two kinds of memory. In addition, older prelinguistic features, retained as paralanguage, were adapted to reinforce the rhetorical powers of verbal artifacts. As gesture and prosody (intonation, amplitude, duration, etc.), they supplied persuasive affect and nuance to speech, as well as formal structures for the expressions of imitative play in narrative, drama, ritual, and song.

In my epilogue, I consider the impact of writing on an oral poetic tradition that might have lasted over 100,000 years. While I can only touch on a few issues, I point out how writing, as an external memory system, made most mnemonic structures unnecessary and prose possible, how the absence of a visible oral performer allowed readers to freely generate mental imagery in response to verbal cues, and how the availability of easily rereadable texts invited writers to create more complex representations of human thoughts and feelings.

I end by reasserting the generally accepted principle that evolution is a cumulative process by which older adaptations are kept in reserve or reused for new purposes. This applies to the successful cognitive traits our nonhuman primate ancestors possessed, as well as those our genus and our own subspecies later developed. As each new stage commenced, and as new means of directing attention appeared, the older means slipped into the background, but, when they did, they continued to perform functions essential to the survival of the species. This means that prelinguistic expressivity was incorporated into spoken language, the oral vitality of which was later incorporated into written texts.

Before you and I enter my paleopoetic time machine, having just outlined this projected mission into prehistory, I want to reaffirm some of the thoughts I began with in this introductory chapter.

By "poetics" I do not refer restrictively to what we literates now call "poetry," i.e., compositions in verse: by using the phrase "verbal artifact" I mean to indicate a much more inclusive cultural *technê*. In my concept of "paleopoetics" I include the skills that prelinguistic humans practiced, skills that, when language evolved, were expressed in verbal structures. It is this repertoire of techniques that, having formed the preliterate imagination, flourished well over fifty thousand years before writing and the literate imagination first emerged a mere five thousand years ago. I have tried to set forth the *idea* of a paleopoetics by first placing it within the domain of a cognitive poetics that incorporates the full implications of the phrase "embodied mind" and that therefore regards cognition as including perception *and* action, language comprehension *and* mental imagery, information processing *and* emotion. At this point in history, the defense of poetry with which I began this chapter can be mounted only by a defense of *poetics*. With this in mind, I will now close this chapter with some brief summary remarks.

As a symbolic system, language constitutes our principal means of sharing information. What intention is to means, so rhetoric is to language. Cognitive rhetoric, the study of the art of speech as hearers (and readers) process it, examines how skilled speakers deploy clusters of words as found tools. Cognitive rhetoric therefore has as its disciplinary basis the science of cognitive linguistics. With that perspective, we may view the work of researchers, such as Ronald Langacker, Leonard Talmy, and the functionalist, Talmy Givón, as supplying the general principles for the more specific inquiries of cognitive rhetoricians, such as George Lakoff and Mark Turner.

Cognitive poetics, as I envisage it, is the study of verbal artifacts as made tools. When we engage these tools and they shift their status from that of objects to that of instruments, they reveal their rhetorical affordances. Since this engagement activates the words, transforming them into the simulations of perceptions, memories, thoughts, and emotions, the verbal artifact is a cognitive tool that can only be understood in reference to the cognitive actions it facilitates.

Can there be a cognitive hermeneutics? Not if by "cognitive" we mean those processes associated with perception, imagination, memory, and other essentially nonverbal representations. Moreover, if a verbal artifact is, by definition, a cognitive tool, it cannot be understood apart from those cognitive processes that activate it and in turn are activated by it. It follows that it ceases to be a tool when it assumes the status of an object, which is precisely how hermeneutics must engage it. The only justification for hermeneutics, after all, is the breakdown of a communicative tool. Let me be clear: hermeneutic theory and the practice of literary interpretation have legitimate functions to perform, but these functions can be termed "cognitive" only in the narrowest definition of that word. As for cognitive poetics, it cannot incorporate hermeneutic aims and perspectives without delegitimizing its own discipline. To assume that it, or any other poetics, must justify its existence by supplementing the work of interpretation is to transpose those two activities, and, as Thoreau once wisely remarked, "The cart before the horse is neither beautiful nor useful."

two
From Dualities to Dyads

Evolutionary forces have shaped organisms to respond to their environments in ways that maximize speed and effect while conserving energy. When possible, therefore, organisms execute multiple simultaneous functions and do so without conscious effort. We, for example, live our lives on two complementary levels, one conscious, the other subconscious. While the conscious may be more eventful and memorable, the subconscious is much busier and more vital, its peripheral and autonomic nervous systems continuously monitoring and adjusting our temperature, pulse rate, and chemical balance. Paired with this subconscious involuntary capacity is our central nervous system, with its surface of activity that we experience as consciousness. Our subconscious state, indicated by a feeling or mood, and our conscious state, interacting with our environment, are each complex processes that can be run concurrently. Linking these two levels, an autonomic-to-voluntary-to-autonomic feedback loop prompts us, from time to time, to choose to modify our immediate surroundings in order to reestablish our optimal internal state. We feel hunger or thirst and search for food or drink. We feel cold or heat and need to find external means to adjust our body temperature. We feel loneliness or fear and seek the help of others.

In this chapter I will be examining some of the dual ways we have of receiving information, processing it, and actively responding to it. From these dualities I will extract what I will call the "dyadic pattern." By exploring the implications of this pattern, I hope to derive principles that we can later apply to our understanding of cognitive poetics. Though I invite the reader to anticipate the application of these concepts to poetics, my focus in this and the next two chapters continues to be upon *non*linguistic activity. In examining the available research on these processes, I will be looking for general features preadaptive to language that may also provide insights into the verbal artifact as a cognitive tool and answers to the ultimate question that cognitive poetics poses: How does such a tool, attached as it is to the faculties of perception, imagination, and memory, actually extend those powers?

Duality

With our two arms, two legs, two eyes, two ears, and all the rest of our binary symmetry, we humans seem anatomically predisposed to dichotomous thinking. Once we can sort all X from all Y, we feel a sense of satisfaction, and, a bit like the god of Genesis after he separated the light from the darkness and the land from the sea, we too pronounce our work "good."

In retrospect, the body–soul and the matter–mind dualities have seemed so right for so very long perhaps because they *were* dualities. More recently, the split-brain discoveries of the 1950s and 1960s, though based on monistic principles, were received with so much enthusiasm, partly at least, because once again a dualistic model seemed vindicated. Of course, Roger Sperry and his student Michael Gazzaniga had revealed new and astonishingly fruitful evidence of bilateral functional differences. That should have been more than enough. However, the popular acceptance of the so-called right-brain/left-brain model and its application to everything from politics and economics to sexuality and child rearing testified to a public appetite for simple, dichotomous solutions to troublingly complex problems.

By the 1990s the popularized "two-brain" version of bi-hemispheric brain physiology had descended to a level of credibility shared by the daily horoscope and the printed slips in fortune cookies. But just as the "two-brain" model faded, psychologists and philosophers of mind began seriously considering a new duality. This was the "two-mind theory," or, to use its less flashy names, the "dual-process" or "dual-systems" theory. Putatively also grounded in neural architecture, this duality was divided not right-to-left but inner-to-outer—i.e., between the older, deeper areas of the primate brain associated with fast, intuitive processes and the newer, specifically human, areas associated with slower, deliberative processes. However, the focus of dual-process theorists was less upon the evolutionary implications of this distinction than upon issues involving judgments (moral and pragmatic), decisions (based or not based on predictable outcomes), and reasoning (rational or biased). One of the pressing questions they posed was this: How does the deliberative mind manage to cope with a partner that trusts its instincts, never questions its own beliefs, and habitually leaps before it looks?

Dual-Process Theory draws upon, and offers a model for, a number of disciplines, e.g., social psychology, cognitive science, robotics, and epistemology and, though the questions it asks are the old ones, this is a twenty-first-century project hoping at long last to find the empirical means to answer them. As is so often the case, when a provocative new theory presents itself in general terms, it inspires a multitude of definers. Table 2.1 is a compilation of those features most frequently ascribed to the two cognitive systems (Frankish, 2010:922).

This table represents a field that, since the mid-1990s, has seen a vigorous exchange of papers, a scrimmage that Jonathan Evans, one of its leading proponents, has endeavored to referee. Surveying all this theorizing, Evans himself has recommended as the best characterization of the two different "minds" the distinction between the heuristic and the analytic. By "heuristic" he means the rapid, intuitive strategy that uses a mix of ad hoc models, concrete examples, and similar experiences to discover solutions to novel problems. By "analytic" he means the slow, deliberative strategy that uses rules to break down a problem into its constituent parts and address them one by one.

In this table, typical of dual-process formulations, the horizontally paired dualities represent logically crisp differentiations, but the vertically differentiated categories tend to be logically fuzzy. For example, the meanings of "fast" and "automatic" overlap: these System 1 processes are fast *because* they are automatic. In System 2, "serial" architecture is a *necessary precondition* for "working memory." With considerable reluctance, Evans came to adopt this broad nomenclature, System 1 and System 2, in summarizing the work of his colleagues, but he himself has preferred terms such as "types" and "clusters" of features (2003, 2008). As he has concluded, these cognitive "systems" are not coherent systems but rather generic types of features (processes, predispositions, strategies, etc.) that may be categorized as either rapid, parallel, and nonconscious or as slow, serial, and conscious, the former features serving a heuristic purpose, the latter an analytic purpose. In referring to the dualities proposed in this theory, I will take the convenient way out and refer to them as S1 and S2.

TABLE 2.1 Features commonly ascribed to the two systems according to Dual-Process Theory

	System 1	System 2
Processes	Fast	Slow
	Automatic	Controlled
	Nonconscious or preconscious	Conscious
	Low effort, high capacity	High effort, low capacity
	Heuristic	Analytic
	Associative	Rule-based
Attitudes	Implicit	Explicit
	Cultural stereotypes	Personal beliefs
	Slow acquisition and change	Fast acquisition and change
	Fast access	Slow access
Content	Actual	Hypothetical
	Concrete	Abstract
	Contextualized	Decontextualized
	Domain-specific	Domain-general
Architecture	A set of systems, modular	A single system
	Parallel	Serial
	Does not use working memory	Uses working memory
Evolution	Evolutionarily old	Evolutionarily recent
	Shared with animals	Unique to humans
	Nonverbal	Language involving
	Serves genetic goals ("short leash" control)	Serves individual goals ("long-leash" control)
Variation	Independent of general intelligence	Linked to general intelligence
	Little variation across cultures and individuals	Variable across cultures and individuals
	Relatively unresponsive to verbal instruction	Responsive to verbal instruction

To the extent that the processes of each pair are distinct from one another, the outcomes they achieve are likely to be different from one another and sometimes incompatible. This has raised questions such as: Do they generate competing solutions that must be adjudicated to avoid the stress of cognitive dissonance? Does the deliberative S2 intervene to inhibit or correct the bias-prone S1? The human genotype, like that of every species, certainly has its maladaptive glitches. The cognitive evolution that produced all those sapient features listed in the right-hand column of table 2.1 did not seamlessly incorporate them into the preexisting left-hand column, nor did it guarantee that S2 would regularly override S1.[1]

While many researchers have tried to establish reductively the independence of each of these two "systems" and assess the rivalry between them, others have recognized the importance of their interaction and the ability of the two "systems," or subsystems within them, to complement one another (Evans, 2003:458; Keren and Schul, 2009:539–41; Darlow and Sloman, 2010:388–89). Independence, after all, does not necessarily preclude collaboration: to the extent that these two dualities can function independently of one another, they have the capacity to do so in parallel. From an evolutionary perspective, this makes sense, since selective pressures would favor adaptations that may be tweaked to reduce systemic conflict.

Before we consider those situations in which features from the left and the right columns come together and actually coordinate one another's performance, we need to recognize that Dual-Process Theory has been formulated to represent the language-enhanced mental equipment of modern *Homo sapiens sapiens*. A Dual-Process Theory that might account for the cognitive skills of our prehuman ancestors (ca. 3 mya) would have to revise its table of features. For example, it would likely attribute to them some S2 features—e.g., "conscious," "high effort, low capacity," "serial," "uses working memory," and "serves individual goals." Early tool-using humans, such as *Homo habilis*, would probably possess all the features of S2 except "rule-based," "hypothetical," "abstract," "single system," "language involving," and "responsive to verbal instruction." It took a long time since *H. habilis* for modern linguistic humans to evolve, at least 2 million years. One would have to be either a creationist or a saltationist to believe that S2 was added to S1 suddenly or in the absence of gradually established preadaptive structures.

At whatever point in time each separate feature listed in the right-hand column emerged, it introduced a new skill into the human genotype, a skill additional to one already in place in the left-hand column (Stanovich and West, 2003:202–3). To borrow the terms of table 2.1, these new skills, "unique to humans," supplemented old skills, "shared with animals." This "shared with animals" feature, we should note, is superadded to *all* the entries in S1. Each and every one of these features, while not uniquely human, was and still is nonetheless a genuine and indispensable property of the human genotype. Dual-process theorists are quite clear about this.

Unfortunately, however, the lists in dual-process tables can be interpreted as either/or contraries. Words like "fast" and "slow," "implicit" and "explicit," "nonverbal" and "language involving" do sound mutually exclusive. Yet here a logical clarity obscures an actual complexity. At every moment, sleeping and waking, the brain is engaged in innumerable neural operations that run at different speeds. Some rely on past experiences, others on an analysis of current circumstances. Some are based on spatial mapping and mental imagery, others on language-mediated thought. Most of these operations the brain can manage concurrently, and, though some may work at cross-purposes, the vast majority of them are collaborative and efficient. It is more useful, therefore, to hypothesize that these opposites are not either/or contraries but rather both/and correlatives. With that in mind, I propose to focus on those features in S1 that interact with and so strongly enhance features in S2 that their duality may be regarded as a two-part unity, a *dyad*.[2]

In addition to those listed in the dual-process literature, there is yet another duality. This is the perception–action duality, which, I would argue, is so coordinated as to qualify as a dyad and so fundamental as to be termed the Master Dyad. Related to other pairs, such as cause and effect, stimulus and response, and input and output, perception and action have been assumed to function peripherally, outside the bounds of central cognition, when the latter is defined restrictively as information processing. In Susan Hurley's caricature of this position, perception and action have been regarded as the two slices of bread enclosing the meaty center of cognition. She called it the "classical sandwich conception," "classical" in the sense of standard and traditional (1998:20–21, 401–12). According to this received belief, perception and action are mutually detached and are the proper purview of physiological, not cognitive, psychology.

To uphold the "classical sandwich conception" one would first have to ignore how mutually dependent perception and action are, ignore the fact that perception, cut off from action, cannot test itself in the object world and becomes the flickering firings of a "brain in a vat," and that motor action, cut off from perception, is little more than an aimless flailing about. According to the Common Coding Theory, the afferent nerves that bring sensory information *to* the brain and the efferent nerves that send motor impulses *from* the brain to the skeletomuscular system communicate directly (Prinz, 1983). This would be one way to explain the speed with which some perceptions are converted into action. The kick of the zebra, the leap of the lion, and the quick shutting of the eye when a small object flits toward it all exemplify the close connectedness of sensory input with motor output (Sperry, 1952; Gibson, 1979; Mandler, 1985).

In this and subsequent chapters, I will have occasion to support Hurley's view that perception and action not only have an interactive relationship but are also each deeply entwined with conscious cognitive processes, including those listed by dual-process theorists. Admittedly, they are not isomorphic with those other paired opposites: neither perception nor action is evolutionarily prior, and each can be "fast" and "slow," "low effort" and "high effort," etc. Their main qualification for being termed dyadic is that these two interactive functions operate simultaneously. How they relate to other cognitive functions will require closer analysis, which I will begin shortly. In the meantime—forgive the analogical leap—if consciousness were not a sandwich, but instead a room, the two elephants in it would be perception and action.

Perception: The Parallel–Serial Dyad and Episodic Consciousness

The connection between perception and action clearly goes both ways. As much as perception directs action, action corrects perception. This is especially true in visuomotor tasks. The eye muscles that adjust for distance, track motion, and shift rapidly from one fixation point to another are, like all muscles, controlled by the brain's motor system, which, when necessary, repositions the two eyes by head movements and locomotion so they can explore the environment from a series of different vantage points.

Sensory organs able to register sounds and patterns of light seem to have first evolved around the onset of the Cambrian explosion (ca. 452 mya). New species proliferated and, under selective pressures, adapted improved means of predation and escape. Along with increased speed, armor, and strength, these new forms evolved new means of locating food and sensing danger in their environment. The prototypes of the visual and auditory systems that we have inherited were next tested during what has been called the Cambrian "arms race."

As highly social animals, our primate forebears (ca. 60–3 mya) needed to do more, however, than perceive food and danger. Their survival also required them to perceive and correctly interpret one another's actions and intentions. As Merlin Donald has proposed (1991), this led to the evolution of what he calls "episodic consciousness," a progressive extension of conscious time from the short-term frame of working memory to an intermediate-term frame. As Donald describes it, an episode is a complex event-perception that can be experienced only if one's brain is able to process a series of related events of different overlapping, therefore parallel, durations and bind them all into one meaningful package.

In his paper "The Slow Process" (2007b), Donald elaborated his earlier episodic hypothesis and its relation to extended online memory. He also discussed how this change may have come about. The origins of the modern mind do not lie in some nature–nurture (or nature–culture) opposition, he wrote, but rather in a "brain-culture symbiosis" (216) brought about through a "brain-culture co-evolution" (217). Our brain had long ago evolved a unique aptitude for interacting with socially distributed cognition as exhibited by the behavior of others. As our human ancestors devised the artifacts of material culture, they also evolved the means to interpret this cultural armamentarium.

Minding one's own business would never have led to human culture. As inheritors of primate social intelligence, early humans survived and prospered by virtue of their curiosity as to the doings of others. "What is he up to? What is he *really* up to? What is she thinking? What does she know about them? How can I learn to do that?" At what point humans put these thoughts into some communicative code, gestural or vocal, we can only conjecture, but it is safe to assume that inquisitive thinking long predated such a code. Accordingly, one long-standing factor that drove the evolution of language was the need to give and get information about one another's neighbors (Dunbar, 1996). In our own twenty-first-century society, electronic social media provide a new means to satisfy an exceedingly old human need.

The ubiquitous theme of epics, folk ballads, dramas, novels, and films has been social interaction with all its rich entanglements, scheming, misunderstandings, and revelations and with all their tragic and comic consequences. Literary Darwinists do have this point right. This fact, Donald remarks, "testifies to our obsession with complex social plots and narratives" (2007b:220). The British anthropologist Robin Dunbar would note that such themes satisfy our primate impulse to share knowledge about third-person others, i.e., to gossip. While endorsing Dunbar's theory of the social origins of language, Donald is also concerned with our capacity to cognize plots and narratives:

> What is the cognitive element, missing in primates, that has enabled human beings to master so complex a social life? One possibility is that apes lack a capacity for the wide temporal integration that is necessary to cope with the intricate plots and sub-plots of human life. The continuous integration of new events into old scenarios, so common in human social cognition, allows the mind to oversee short-term events... from a deeper *background* vantage point, while bracketing the fast moving events in the *foreground*, and placing them in an accurate context. (220, emphasis added)

Being easier to test, the fast-moving "sensorimotor foreground" has been the focus of cognitive research. Perhaps that explains why so little is known about the slower, wider time frame (the intermediate time zone) in which social interactions occur. Empirical, neurophysiological evidence as to how this process fits into our cognitive architecture has yet to be established (Donald, personal communication, Oct. 4, 2010). Nevertheless, it is this " 'slow process' in the brain that is uniquely human":

> In effect, the hypothesized "slow process" is a vastly extended working memory system that serves as the overseer of human mental life, and is the deepest layer of the mind. This is the intermediate-term governor of human mental life, the deep *background* process that shapes our cognitive agendas over the longer run, while maintaining oversight over the *foreground* of mental activity that occurs closer to the sensory surface. (Donald, 2007b:220, emphasis added)

The dyadic nature of this "slow process," this "extended working memory," lies in the fact that this cognitive skill requires a close interaction of working memory, which narrowly focuses on and binds successive events, with a "deep background process" that generates a broad context for these events. It could not be accomplished otherwise.

In the two passages above, I have added italics to draw attention to the gestalt implications of Donald's model. Episodic consciousness, as it continued to evolve in the brain of early *Homo*, replicated that long-established manner of organizing a perceptual array into narrowly focused, clearly demarcated items (figures), while continuing to be aware of a broadly scanned, less detailed field of surrounding items (ground). As Donald's choice of words suggests, the moment-to-moment events in an episode constitute a relatively fast-moving succession of foregrounded figures, while the already registered events remain in the (back)ground where they continue to supply a meaningful context for each subsequent figural event. In other words, within an episode the accumulated events, together with our assessment of them, frame each new event as a visual ground frames a visual figure.

To interpret an episode this way, we must consider how figure–ground principles apply not only to the visual perception of spatial arrays but also to the brain's cognition of temporal events, successions of actions and multimodal perceptions connected in both short-term and intermediate-term working memory. An episode is not merely an action: it is a temporal gestalt that, like a topological entity, may be stretched or condensed without losing its essential figure–ground structure. Episodic consciousness can exist in time, only because our brains are capable of processing perception events both as successive foregrounded figures and as a deeper background of already interpreted actions.

When we reflect later upon an extended episode, e.g., an hour's interaction with neighbors recollected the following day, our memory reassembles it in a similar figure–ground format. Endel Tulving (1983) called this capacity "episodic memory," because he recognized that the content we retrieve from this long-term store is composed of separate "episodes," event clusters, each grounded in less defined, largely forgotten intervals. Memory thus arranges our past, as well as our present, according to the figure–ground duality, a quasi-spatial temporal design that narrative has universally adopted.

Action and the Anatomy of Multitasking

Human action is made possible through anatomical dualities that long preceded the figure–ground perceptual duality and may indeed have furnished the template for it. The anatomical design that we share with all vertebrates reveals structural symmetry together with a tendency toward *functional asymmetry* (Corballis, 1993:80–108). This means that, while some symmetrically paired organs normally operate similarly, they can, when called upon, execute different tasks. For example, the front and hind legs of quadrupeds show little differentiation in straight-ahead running, but the hind legs of some have been adapted for quick swerves or backward kicking, while the front legs of others have been specialized for grasping. To be efficiently performed, such asymmetrical functions must be done simultaneously: the hind legs of the lion must continue to propel it forward while the front legs leave the ground, reach forward, and converge; the hind legs of the zebra must kick back while the front legs momentarily plant themselves. In short, these asymmetrical actions of otherwise symmetrical limbs must become automatic and operate simultaneously.

When a species comes under stress, new traits, including those that require simultaneous actions, may prove successful enough to be passed on to future generations. But new forms of skeletomuscular multitasking are difficult to acquire when the neural wiring that must control both component actions has heretofore been shared within the brain and central nervous system. Unless these neural pathways are sufficiently dissociated, a phenomenon known as *interference* occurs: one task inhibits or retards the other, resulting in either an inefficient performance of both or a need to queue the tasks serially.

It could not have been easy for hominid apes (chimpanzees, bonobos, gorillas, gibbons, and orangutans) to become the resourceful creatures they now appear to be. It took them some 15 million years to distinguish their skills from those of bushbabies and tarsiers. It was, for example, quite an achievement when, instead of employing all four legs to run and climb, they reserved one forelimb to carry an object, such as food or young, and came to move about tripedally. It marked an even higher level of manual coordination when instead of needing both arms to lift a rock, they found they could use one to raise it and the other to search for grubs beneath it, or when instead of using both hands to swing from a tree branch, they learned to use one to hook on and the other to pluck fruit (Kelly, 2001). Tripedalism may also have set primates on the road to functional specialization both in the hands and in the two hemispheres of the brain (Fagot and Vauclair, 1991).

As our prehuman ancestors became fully bipedal (4–3 mya), their arms assumed a new swinging function that helped them while walking and running. Each arm would swing forward, then backward, opposite to the placement of the ipsilateral foot. According to one recent study (Pontzer et al., 2009), contrary to common belief, arm swinging adds nothing to the momentum of runners. Instead, its purpose is to maintain upper-torso balance and stabilize the head so that the eyes can hold their focus on an object in front of them, e.g., a fleeing animal. As an aid to vision, the arms counter the lateral torque of the hips that would otherwise jog a focalized object from left to right, falsely indicating a horizontal change of direction. If that object were a game animal, this confusion would diminish a pursuer's chances of running it down. Bipedal locomotion is evidently one part of a more complex motor routine, one that coordinates legs, arms, and eyes in actions aimed at achieving a specific goal.

In the course of becoming bipedal, the hands of our hominid ancestors also gradually changed. The fingers shortened and straightened, the thumb grew stronger, longer, and more able to press down upon the tips of the fingers. Over time, they found more and more things to do with their hands, such as tool use and gesturing. But to say that walking upright left their hands free for these other purposes, while undoubtedly true, risks oversimplifying a complex evolutionary process. As Richard Young (2003) points out, an upright posture also enhances the effectiveness of throwing and clubbing, two biomechanical skills that would have literally given their users the "upper hand" in hunting and warfare. (Just imagine hurling a stone or swinging a stick from a seated or a three-point stance.) The evolution of bipedalism may therefore have been driven by the selective advantage not merely of pursuing prey but also of throwing and clubbing while running. Here again we have an instance of coordinated multitasking, a synergy of legs and arms used in order to strike down an object targeted by the eyes.

Though we commonly speak of the "dominant" hand (usually the right) and the "nondominant" hand (usually the left), we should not think of this difference in terms of preference and neglect. For most bimanual tasks, each hand has its own special function. Two examples come readily to mind (and here I will assume right-handedness as the norm): we use the left hand to rake up a mass of objects and the right hand to select individual items, and then we use the left to hold a selected item in place while the right does something to it; scooping or holding objects, the fingers of the left hand tend to work automatically, often in the peripheral visual field, while the right hand performs a careful series of actions, monitored by focal attention. The functions of our two hands, right and left, can thus be mapped onto the two columns of the dual-process table of features (table 2.1). Like our left hand, the features in the left-hand column oversee broad swaths of data that the features in the right-hand column sort out and selectively attend to.

Rather than using the terms "nondominant" and "dominant," the French neuroscientist Yves Guiard (1987) has proposed we refer to manual actions as "macrometric" and "micrometric," respectively ("metric," as in spatial scale and measurement of time). He observed that, when both hands function in parallel, the left regularly initiates the action, followed by the right, and further suggested that every human action, be it perceptual, motor, or intellectual, tends to proceed from the macro to the micro level. In figure–ground terms, the macro level is the ground that is scanned first before a figure is located on the micro level. As I will show in a later chapter, Guiard's bimanual theory has intriguing implications for our understanding of sentence structure.

The functional asymmetry of right and left hands leads us to consider the functional differences displayed in the prehension and manipulation of objects. In 1956 the British anatomist and primatologist John Russell Napier first introduced the terms "power grip" and "precision grip." In the power grip, used while grasping a cylinder, such as a stick or a club, all four fingers plus the opposable thumb wrap around the object and press it firmly against the palm. In the precision grip, used while holding a relatively small spheroid or delicate object, only the tips of the first two fingers and thumb supply the pressure. As Guiard might put it, the right hand, best fitted for micrometric tasks in over 90% of humans, specializes in using the precision grip, while the left often applies the macrometric power grip to hold an object in place. Either hand, of course, can use either grip, but neither hand can manage a third kind of grip, Napier said. Though he allowed for slight variations, these two are the standard grips, and it was therefore incorrect, as many physiologists had assumed, that every object called for its own proper grip.

Napier's work reached a wider public when he was called in to identify fossil remains found by Louis Leakey and his colleagues in the Olduvai Gorge in Tanzania between 1959 and 1963. Based on his examination of finger bones, Napier agreed that these remains belonged to a true human, not an Australopithecine, and that this individual was capable of both the power and precision grips. This, and the fact that crude stone tools were found close by, led him and the others to conclude that this was a member of the earliest tool-making species, which came to be called *Homo habilis*, or "handy man,"[3] a finding that intensified the ongoing debate on the relation of tool use to the evolution of the human hand (Leakey et al., 1964; MacKenzie and Iberall, 1994; Marzke and Marzke, 2000).

I referred earlier to Richard Young's hypothesis that the throwing and clubbing that coevolved with bipedalism among prehuman hominids required a very precise muscular coordination of legs, pelvis, upper torso, arms, and eyes. He also, of course, incorporated into his thinking Napier's analysis of grip: throwing used the careful control and release of precision grip, whereas one-handed or two-handed clubbing used the power grip. While most paleoanthropologists had long concluded that the human hand had evolved alongside tool making, what Young proposed was that these skills evolved when *found* tools were first used, which was long before *made* tools were crafted, or as he put it: "Adaptation for improved throwing and clubbing would have *pre-adapted* the hand for stone knapping" (Young, 2003:171; see also Calvin, 2004:128–30). We are once again reminded that every innovation has had a long preparatory past.

Natural selection often couples long-term survival benefits with short-term pleasurable rewards. The exhilaration associated with successful multitasking underscores its evolutionary importance. Most of the sports we enjoy as participants and spectators require a combination of alert perception and precisely coordinated skeletomuscular action. But not just any sort of coordination will do. The multitasking displayed by an expert juggler requires talent, practice, and athleticism, but we do not place it in the same category as a sports performance. It does not feature running coordinated with upper-body movement. Though otherwise diverse, our modern sports of basketball and American football combine running and throwing; tennis and hockey combine running and clubbing; and cricket and baseball involve the coordination of all three actions.[4] The reason we employ and admire these nonproductive skills may be that they connect us to a past, still present in our neural circuitry, a past in which pursuing, throwing, and clubbing were the prime skills that permitted our genus to compete successfully with other carnivores and gave gracile *Homo sapiens* tactical advantages over all other human species.

The artifact evidence indicates that Oldowan craftsmen had been predominantly, but not overwhelmingly, right-handed (Schick and Toth, 1994). As paleoanthropologists have determined, the knapper's left hand grasped the object to be struck in the power grip, while holding the hammer stone in a firm, right-handed precision grip. Cushioned by the palm, the hammer stone was kept in position by thumb, index, and middle finger, all three lying along the axis of the forearm. Rather than swing it, the knapper forcefully jabbed the hammer stone at precise places on the core stone.

We, too, favor the right hand for tool use, but it is interesting to note that when we use our own standard hammering device, we do not use the same grips they used. When we prepare to hammer a nail, we hold the nail in place between our extended thumb and index finger in the precision grip, while we grasp the haft of the hammer in the power grip. The risk of striking our thumb instead of the nail comes from the fact that the power grip requires a more complex coordination of arm and shoulder muscles, whereas the precision grip, even for the nondominant hand, is simpler, finer-tuned, and therefore better controlled. If, as this reversal indicates, the precision and power grips, together with the actions associated with each, are not firmly linked with the dominant and nondominant hands, respectively, this dyadic division of labor is evidently not hardwired in human anatomy. Is it possible that this integrated duality, this dyad, is actually a *pattern*, an evolutionary strategy for accommodating novel adaptations within older systems? If so, we should expect to find this strategic pattern replicated in other functions. As I proceed in the next section to examine the central cognitive activity of information processing, I will be looking for further instances of this pattern.

The importance of the hand in human evolution cannot be overstated. The division of labor between the left and right hands gave humans a powerful new means of dual tasking and, through the extension of the grasped tool, led our genus from biological to cultural evolution. The next step, however, was to accelerate the evolutionary pace. It came when our ancestors succeeded in taking the duality of the power and precision grips and combined both these grips *into a single hand*. In the "combined grip," as Napier named it, the tips of the thumb and index finger come together in a precision grip to manipulate a part of an object while the last three fingers hold the rest of it in place by using the last three fingers to wrap around it in a power grip. In this way a single hand can, in effect, perform two tasks simultaneously. As Napier remarks (1980:119), we use this when we tie two ends of a rope together: the index and thumb of each hand holds an end in a precision grip while the other three digits grasp the bight in a power grip. This combined grip made it possible to braid fibers into cords with which to sew leather garments and shelters, thereby making it possible for humans to migrate into colder climates in their pursuit of game.[5] By coordinating in a single hand the broad, macrometric features of the power grip with the narrow, micrometric features of the precision grip, the combined grip further illustrates the dyadic pattern and its crucial role in human evolution.

Information Processing: The Parallel and Serial Modes

If the mind is a sandwich, to take a final bite of Susan Hurley's satirical analogy, and perception and action are the enclosing slices, then information processing is what forms the meaty cognitive center. For this center, the most important duality is represented by those two modes of organizing data and executing action, the parallel and the serial. Briefly defined, the *parallel* mode is the organization of simultaneous streams of data as a single event, while the *serial* mode is the organization of data as a succession of events.[6] Despite their differences, we have seen how these two modes are also profoundly interdependent and complementary in their functions (Treisman and Gelade, 1980; Cave and Wolfe, 1990). While the parallel mode is, by definition, a coordinated action of different elements, the parallel mode *and* the serial mode constitute a very powerful dyad whenever they act together to accomplish a task.

Merlin Donald's first two stages were made possible by advances in these two processing modes. The episodic stage marked an increase in the capacity of the primate brain to parallel-process perceptions and derive social information from them. The mimetic stage marked an increasing ability to execute actions in a step-by-step serial fashion. When language eventually emerged in the mythic stage, it adapted these two already established modes to build its own structures. For example, the sounds that form words are serially produced by a speaker and must be serially processed by the hearer. At the same time, however, the speaker and hearer must also be able to integrate the meanings of whole phrases and sentences by organizing them through parallel processing. The "at the same time" phrase is appropriate because our parallel and serial functions, in speech and in many other skillful routines, have indeed evolved to run at the same time as a dyad.

Much of what I have just described as multitasking and bimanual coordination has been parallel output. From an evolutionary perspective, the parallel mode is as old as unicellular life forms. It is the default mode: in any emergency, living things opt for the parallel processing of incoming perceptions and of outgoing actions. When the serial mode later emerged, it became as narrow as the parallel was broad, its attention to incoming information and outgoing action as sharply focused as parallel attention was diffusely distributed. The serial mode would become increasingly important in human evolution in the mimetic stage when our ancestors began to use and, later, manufacture tools by which to modify their environment. This, in turn, modified the brain itself, for, as we have come to learn, the brain's left hemisphere became specialized for serial activities, for "stringing things together," as William Calvin aptly put it (1993:231–34). Since it is a skill to which humans owe much of their tactical superiority to other animal species, I will now focus my attention on serial information processing.

As the serial action of pedal locomotion is fundamental to all warm-blooded land animals, mental mapping is fundamental to foragers. Being the principle upon which these maps are constructed, it was the serial mode that our ancestors relied on to find their way to food and water. At some later point they probably learned to extend metaphorically this notion of a (walk)way to that of a (work)way, understanding that every purposeful action has a goal and that a goal is unthinkable except as the end point of some step-by-step progression. We still commonly speak of a "way" when referring to a prescribed set of actions. We look over one another's shoulders to give *directions* that do not imply actual traveling: "Don't do it *that* way—do it *this* way." The words "process," "procedure," "method," "routine," "course (of action)," and "step" (as in "step one, step two . . .") were originally all versions of this one metaphor.

Before proceeding further, it might be useful to distinguish *successive* from *sequential* serial events. In a succession of events, there is no foreordained order: one event simply happens subsequent in time to another event. We use the "way" metaphor in the successive sense when we say, "That's just the *way* it is" or "He has an interesting *way* of looking at the world." Here we want to draw attention to certain inherent, but perhaps not obvious, properties of a condition or a person that tend to reappear consistently but in no particular order. In a sequence, however, one event necessarily follows another event in a predetermined order. It is significant that serial processing of the *successive* variety is an evolutionarily older feature that, when dual-process lists are drawn up, rightly belongs in the S_1 column.

Like other animals, we use successive perception in order to pick out significant sounds and sights. As we take in the soundscape, we listen for patterns that will indicate the source and direction of particular sounds and then find ourselves able to filter out all other frequencies (the "cocktail party effect" preceded mixed drinks by many millions of years). Our visual system can also use successive seriality. When we survey a landscape, for example, we move our focus successively over whatever area seems of interest to us, our gaze darting about the whole array, returning repeatedly to fixate on particular objects. Serial sensory input has evolved to be successive by default. Consequently, even when we hear speech or see writing, we process these data as serially successive, then we scan for meaningful patterns and construe these as they gradually unfold. The words may indeed unfold in sequential order, but we do not comprehend their meaning in a rigidly linear fashion. This is quite different from how we manage a strict sequence of items, e.g., the digits in a combination lock, the alphanumeric password that lets us access a Web site, or the memorized words of a song or poem. For these, strict sequentiality is essential.

Conscious, intentional actions, such as foraging, also involve information that the brain must process. When we walk about looking for food, whether in a forest or in a supermarket, the succession of our actions is no longer randomly exploratory. Though not a task with a prescribed sequential order, a search is a methodical behavior that entails (1) knowing what we are looking for, i.e., having a mental image of the search item that can be matched up with appropriate percepts in our otherwise distractor-filled visual field, and (2) knowing not to reexamine locations we have already searched in vain. The latter skill, which is called "inhibition of return" (IOR), is a hardwired trait in primates, human as well as nonhuman, that works by slowing the reaction time we devote to already viewed objects, thereby accelerating our shifts of attention to novel objects. Search thus becomes a serial process of elimination by which we can safely ignore already examined locations, thus increasing our chances of finding our objectives in the fewer remaining places (Posner et al., 1985; Klein et al., 2001). In short, search is a successive series of actions, but one that is goal oriented, progressive, and nonrecursive.

Now, suppose we return to that place, just two paragraphs ago, from which we were gazing out at a landscape. If we choose to move toward one object—say, a nut tree—we now enter into a more or less sequential motor routine. In traversing the distance between our first position and that tree, we must do some improvising to navigate around and over whatever obstacles lie in our way. Before we can shake loose the nuts that hang beyond our reach, we will need to find a stout stick of an appropriate length. At the tree, we grasp the stick, raise it toward particular nut-laden boughs, and then strike with it. If we want to taste our find then and there, we will need to locate two stones, one to serve as a hammer, the other as an anvil. But if we think we might share these nuts with others or husk and save them, our goal will extend spatially from that tree to our home and temporally from this day to several months in the future. Sequential planning of this sort now entails social responsibility and deferred gratification.

The serial mode has inner, as well as outer, cognitive functions. One of these is the way we recall an episode from our past. As Endel Tulving (1983) defined it, episodic memory retrieval is an effortful replaying of some experience from our past, set in a specific place and time. The fact that the percepts are impossible to access in their original parallel richness means that they can be reconstituted only through a sequential series of, typically, visual images. Not only can we simulate perception using the serial mode, we can also simulate action by using a process that Merlin Donald has called "kinematic imagination" (1999:142–43). This sort of visualization lets us reinforce the neural connections in the brain associated with successful motor sequences, such as musical and sports performances.

The kinematic imagination can apparently improve a person's motor skills in preparation for later serial actions, but can this same imagination respond to online perceptual cues? Researchers into the "mirror neuron system" say yes and, since the mid-1990s, have published findings that have profound implications for our understanding not only of the serial mode but also of the evolutionary origins of language and its artifacts. Testing the neuronal reactions of macaque monkeys to the sight of a human or another macaque grasping, holding, tearing, or otherwise manipulating an object, Giacomo Rizzolatti and his colleagues at the University of Parma discovered that certain cells in one area of the premotor cortex would fire and that these were the same cells that would also fire when that observing animal itself performed such actions (Gallese et al., 1996). If he and his colleagues are correct, this neuron-based dyad of perception and action makes imitation possible, the social behavior largely responsible for our evolutionary success. The prerequisite for social consciousness, as Susan Hurley (2005a) saw it, was the ability of humans to enact within the brain the perceived intentional actions of others—in effect, to lock on to and share one another's neural circuitry:

> The intersubjectivity characteristic of human beings, their distinctive capacity to understand and empathize with one another, is enabled as a specialization of enactive perception: I perceive your action enactively, in a way that immediately engages my own potential similar action, thus enabling me to understand, or to imitate, your action. Shared processing of the actions of other and self is a special aspect of the shared processing of perception and action. (192)[7]

Year by year, research has continued to supply fresh evidence to support the hypothesis that humans do indeed possess their own mirror neuron system, which has not only enabled them to learn and transmit motor skills such as tool making and tool use but also laid down the neural mechanisms out of which language evolved. This latter claim draws strength from the fact that the area identified as the site of mirror neurons partly overlaps Broca's area, the location where human speech articulation is centered. As has long been known, in the prefrontal motor cortex the neurons that control facial and mouth muscles lie adjacent to those that control arm and hand muscles. "The convergence of the empirical data is impressive and suggests shared neural structures for imitation and language," writes Marco Iacoboni. "If Broca's area has an essential role in imitation, then it must be concluded that this area is not exclusively dedicated to language processing. It also suggests an evolutionary continuity between action recognition, imitation, and language" (2005:90–91).

The leading advocates of this hypothesis, such as Rizzolatti, Gallese, Arbib, and Iacoboni, do not claim that humans passed from motor mirroring to vocal language in one mighty leap. The close proximity of Broca's area to the mirror neuron location in nonhuman primates suggests that human language may have evolved first as a system of manual gestures during a period that may have spanned the 2-million-plus years from the onset of tool making to the Upper (Late) Paleolithic. During this period, humans may have gradually developed manual and orofacial signing into a system supplied with a usable lexicon and a rudimentary syntax, a possibility I will explore in chapter 5.

Be it signed, spoken, or written, language happens as a segmented series of events, a stringing together of units. In spoken language this is a series of phonemes that merge into syllables that in turn merge into individual words. These words, available to be serially articulated and serially received, comprise the *lexicon* of a given linguistic community. *Syntax*, as a set of hierarchical rules, allows words to interrelate on the basis of causality, position (below, above, in, out, etc.), time (after, before, during, etc.), similarity, contiguity, and other associational linkages. Just as lexicon, when actualized in discourse, is a serial function of language, syntax permits words to function in parallel. Together, lexicon and syntax further demonstrate the dyadic complementarity of the two modes.

The mechanisms that subserve the serial mode have made human cultural evolution possible by bringing our ancestors to the mimetic stage and preparing them to transit, some 2 million years later, to the mythic stage. But, again, we must keep in mind that the major modifications to our central nervous system associated with these stages are additive, not substitutive. The continuing parallel and the emerging serial have become two complementary modes, a dyad that eventually supplied the creative momentum responsible for forging that prime tool of socially distributed consciousness, language.

The Dyadic Pattern

The multitasking that I touched on earlier, such as using different parts of the body to accomplish different tasks at the same time, was *successful* multitasking. But, as we all know, multitasking is not always successful. This is especially the case when two serial processes vie for that narrow band of focal attention. For example, using a cell phone or text messaging, while at the same time operating a vehicle, creates cognitive interference, leading to a breakdown in multitasking with potentially catastrophic consequences. In less risky circumstances two serial tasks may indeed be performed in parallel, though not at the same level of alertness. To apply the dyadic model of figure–ground perception: one of the two serial actions may be safely relegated to peripheral attention, while the other is centrally focalized, a division of labor to which we owe much of our success as a species.

In table 2.2, I apply what I have called the "dyadic pattern" to a number of cognitive functions. I begin with the pattern itself, a relationship between broad (peripheral, diffuse) awareness and narrow (centralized, focused) attention. "Awareness" characterizes the comprehensive, holistic nature of this preattentive level. "Attention" highlights the concentrated nature of its paired opposite. This pattern represents two metacognitive modes of monitoring and reflecting upon information and then choosing and executing appropriate responses. Each component of this dyadic pattern is linked to perception through the figure–ground dyad and to action through the parallel–serial dyad. Broad awareness is supported by (back)ground sensory input and by parallel motor output. Narrow attention is supported by complete figures detached from their ambient settings and by serial movements. The strong presence of perception and action in this table emends what I take to be a deficiency in Dual-Process Theory: perception and action cannot be sealed off from one another, since both are essential to central, cognitive information processing.

The activities listed below this bipartite division include functions that usually operate on the periphery of consciousness, brain processes such as instinct, emotion, dreaming, and mirror-neuron responses. I also include a number of classically "cognitive" functions, such as memory (short-term, intermediate, semantic, episodic, and procedural), imitative learning, and mental imaging. I do not include language-associated functions, such as categorization, reasoning, logical decision making, and problem solving, because my focus now is upon the cognitive functions preadaptive to language. My division between "outer" and "inner" assumes a level of self-reflexiveness well within the capacity of the primate brain and is, moreover, a useful distinction, as I hope to demonstrate.

I have referred to perception and action as forming the superordinate dyadic pattern, the "Master Dyad." The broad component of this pair is perception, and by that I mean *all* our perceptual systems—vision foremost, but also hearing touch, taste, smell, proprioception, kinesthesis,

TABLE 2.2 Cognitive processes as expressions of the dyadic pattern

DYADIC PATTERN:	1. BROAD AWARENESS perception: *ground* action: *parallel process*	2. NARROW ATTENTION perception: *figure* action: *serial process*
A. OUTER: perception	• Concurrent, preattentive, *multisensory input* from exteroceptors (sight, hearing, smell, etc.); the formation of a perceptual ground.	• *Single-sense attention* to successive input stored in short-term working memory; the formation of perceptual figures.
action	• Use of functionally asymmetrical organs and limbs in *multitasking*; complex tasks stored in procedural memory. • *Episodic awareness*: unified sets of successive events, experienced through extended working memory ("intermediate term memory").	• Oculomotor and locomotor *search* for targeted objects. • *Mimetic skills*: sequential, focused actions and routines, e.g., tool making, tool use, and other manufacture (stored in procedural memory).
B. INNER: interoception and simulated perception	• *Interoceptive awareness*: somato-sensation, proprioception and affect (emotions, "feelings," "moods," and other subjective states). • Automatic retrieval of mental images from *semantic memory* (later enhanced by language).	• *Successive images*: waking visualizations, daydreaming, and REM sleep). • Effortful retrieval of predominantly visual data from *episodic memory*.
instinct and simulated action	• Instincts.	• Online motor simulation via mirror neuron system. • Offline rehearsal via kinematic imagination of sequential routines (stored in procedural memory).

and the rest. All the sensory modalities are there to monitor our environment and, working in parallel, to cast a broad net of awareness about us. When any one, or several, of them receives a particularly strong or recognizably significant signal, we turn our narrow attention to its source and, if necessary, take action. Action is also supported by narrowly focused attention. While this often involves multitasking, action is goal-driven and commits the entire body to a set of serial procedures, which are, all the way to completion, monitored by broad perceptual awareness. In table 2.2 and the discussion that follows it, we will need to distinguish separate dyadic patterns within this overarching Master Dyad.

Now for some brief comments on the text within the table.

1A. Broad Awareness: Its Outer Functions

These, first of all, involve a sensory openness to sounds, sights, smells, etc. The ambience they register is not the "great blooming, buzzing confusion," as William James (1890/1950:462) characterized a baby's early perception of the world, but rather a tempered medley of sense impressions, picked up preattentively in the periphery of consciousness. These form the array from which single-sense attention (column 2) selects and identifies figural entities. Barry Dainton used a walk through a park to speak of these outer perceptual functions. You see a shrub and wonder what it is. A child begins to cry. You hear a birdcall. Each time that you focus on one thing, everything else shifts to the periphery of your attention. "The point I want to get across is that the overall experience here is *unified*.... The phenomenal background is not just a constant presence in ordinary experience, it is a unified presence" (Dainton, 2000, author's emphasis).[8]

Dainton's relaxed walk in the park reveals the ever-present ground of perceptual consciousness, which in normal perception is dyadically coupled with a series of narrowly enclosed figures. It is only with an extraordinary effort that we can attend to this ground without locating figures within it. It is easier to do so when the visual array is reduced to an undifferentiated field, a ganzfeld—e.g., complete darkness, a fog, or the replicated pattern of waves across a calm sea (Collins, 1991a:156–60).

Now for the outer *actions* monitored by broad awareness: these are characterized by the evolved capacity of an animal to use two or more effectors to accomplish some single goal outside its body. This, as I earlier observed, became possible because the structural symmetry of vertebrate anatomy allowed for the differentiated functions of, for example, back and front limbs, right and left limbs, and various digits. These voluntary maneuvers could be used in complexly parallel, extended activities, the successive details of which could be bound together and recognized as constituting whole, ongoing episodes, such as hunts, foraging expeditions, mating, and social play.

We humans also have each acquired and stored in long-term procedural memory a repertoire of variable routines (skills). Consider for a moment the way we use a complex skill, such as driving a stick-shift auto. While one foot is pressing down the clutch, one hand moves the stick; at the same time the other hand is controlling the wheel, the other foot is regulating the gas, and the eyes are all the while scanning the road ahead, to the sides, and to the rear. The word "procedure" implies a series of separate actions, as in "There's a whole step-by-step procedure you have to learn." But once a multitasking skill gets stored in procedural memory, it can be activated and performed in parallel: to a driver who has learned the stick-shift routine, the human process is fully "automatic."

1B. Broad Awareness: Its Inner Functions

These quite logically begin with interoception (internal perception), the largely preconscious monitoring of somatosensory states. This is the information we draw upon when we want to reply seriously to the questions "How are you?" or "How are you feeling?" They also include proprioception, our awareness of our body—arms, legs, fingers, neck, etc.—in space.

When we consider the broad awareness of interoception in relation to *action*, we need to look at the ways the body has to mobilize its systems to act in concert. Some of the work of preparing for action is entrusted to those automatic mechanisms, generally termed "instincts," that enable the body to react to situations with, for example, flight-or-fight or sexual readiness responses. In addition to instincts, many of which are pre-mammalian in origin, mammals evolved a number of "emotions," which each species has customized to fit its biological niche. The fact that emotions entail a longer reaction time than instincts suggests that this time could be used to appraise an emergent situation and choose an appropriate course of action.[9] Emotions, like other affective episodes, may sometimes seem "free-floating," but they are usually triggered by objects and events processed outwardly and focused upon with narrow attention.

Here the figure–ground dyad operates: a narrowly defined figure appears to provoke a diffuse inner ground of affective impulses.

The study of human emotions has been a difficult and contentious field. These affective states are multilevel, interrelated, and labile. By the time one is aware that one is undergoing a given emotion, many neural and visceral processes are already under way, so to describe such a process is like setting up one's easel and trying to paint a sunset. There are reasons why emotions have to be complex. If, as is often the case, a sudden, consequential change in the environment has prompted the body to mobilize its motor systems, it must do so rapidly, i.e., in parallel. On the other hand, the fact that an emotion may be the complex expression of multiple parallel adjustments does not mean that it is necessarily of short duration. Inner, as well as outer, parallel processes can extend over time to inform whole episodes. Moreover, some animals, humans included, can also voluntarily imitate emotions, or at least the outward, "ritualized" appearances associated with them, and can use these displays to signify ongoing attitudes (cf. a dog's bared teeth, a human's scowl). Needless to say, the role of emotion in language and in verbal artifacts is an issue I will need to revisit in later chapters.

Another important inner sensory function involves stored representations of objects. When we think, we often do so in visual images, as our prelinguistic ancestors no doubt also did. The stored images we access from their "places" in subconscious or preconscious storage seem co-present to one another like the objects that coexist in our visual periphery. To visualize any single stored image, we must select it in an introspective act that simulates the way we select a visual figure from other co-present items in its visual ground.

Are these whole images "ideas" either in the Platonic or the Lockean sense of that word? Or are they stored as separate features that are then aroused by the perception or thought of some entity? Connectionist theory, which regards cognition as an activation of deep neural networks distributed throughout the brain, favors the latter explanation. As Lawrence Barsalou (2008) has argued, there are no fixed representations in the brain. Instead, we carry about with us the basic constituent features, image schemas of, for example, "bird" (wings, feathers, beak, and flight) and, on a finer level of specificity, the features of "crow," "duck," "hawk," etc., and, based on our perceptual experience, we select the appropriate features to distinguish one bird from all the others.

These context-free features are the makings of an "iconicon" of prototypes, a sort of field guide to our human habitat. Long before humans came to communicate thought through gesture and voice, this imaginal filing system must have been fully in place, allowing them to recognize objects in general terms of edibility, danger, and other affordances. Long before language, as a symbolic code, permitted them to organize and share this inner sensory iconicon, our remote ancestors each had access to his or her own semantic memory archive. Note: the term "semantic memory," as it is used by neurologists and cognitive scientists, is not restricted to users of linguistic codes but is applied to all animals that have learned to recognize the distinctive features of objects. When language did evolve, the iconicon was joined by a lexicon, and thereafter humans could use words to accelerate the cueing of mental images (Paivio, 2007).

Instinct manifests as an innate, unlearned response to some external stimulus, but it is less an overt action than it is an inner preparation for that action. Unlike an emotion, an instinct (e.g., fight, flight, or freeze) is an automatic, all-or-nothing reaction that has individual survival as its principal purpose. The term, which had been long used in reference to nonhuman animals, was applied to humans from the mid-nineteenth century to the mid-twentieth. It has since lost currency among psychologists except for evolutionary psychologists, who claim that instincts still lie at the source of human behaviors. To whatever extent that claim is valid, it seems likely that the actions of prelinguistic humans were strongly influenced by impulsive, stereotyped routines.

2A. Narrow Attention: Its Outer Functions

Perception, narrowly focused, tends to concentrate attention on one sensory aspect at a time, e.g., appearance, sound, or smell. This it accomplishes by actively attending to successive bits or packets of information. As I suggested earlier, this function scans for meaningful patterns, e.g., the tread of an identifiable animal, the shape of hoof prints, the directionality of a whiff of smoke. Though it can pick out sequences (George Miller's seven units plus or minus two, 1956), it is adapted to draw inferences from a random succession of indexical signs. Its capacity to focus narrowly on a series of details and pick out meaningful wholes from a sensory array produces the figure in figure–ground perception. While this is principally a visual phenomenon, it is not limited to that modality.

Narrow attention in the outer action category facilitated three important early human behaviors—gathering, tool use, and hunting. Gathering and hunting depend, of course, on perceptual skills (successive scanning and figure selection), but these in turn depend on the searcher's oculomotor system and active movement through the environment. As we have already seen, searching for food (roots, fruit, insects, etc.) became faster and more energy-efficient thanks to "inhibition of return," an adaptation that, by eliminating previously searched locations, may have prepared the hominid mind for sequential processing. With tool making and the skills that developed from it in the Paleolithic era, sequential seriality fully emerged and, with it, a progression of skills: fire making, large game hunting, food preparation, shelter construction, and the stitching of leather garments. Stored internally in procedural memory, a skillful sequence was a "stringing together of things" that, like the artifacts it produced, could be used over and over again and, transmitted through imitation, could become a socially replicable technology.

2B. Narrow Attention: Its Inner Functions

To begin with its inner perceptual functions: these simulate perceived episodes, remembered, fantasized, or dreamt. Mental imaging differs from perception in quality and speed of representation. The inwardly imaging brain can never achieve the same speed of processing it attains while encountering outward events. Unlike semantic memory, which is our inner store of context-free knowledge, episodic memory, our store of space- and time-contextualized personal experiences, takes time and effort to access. This is consistent with the principle that associates difficulty with a narrowing ("bottlenecking") of information flow, a condition that necessitates serial processing (Broadbent, 1958; Posner, 1978; Pashler, 1999).

When we improvise an imaginary scenario or recall some lived episode, we can do so only segment by segment, frame by frame. Suppose, for example, instead of simply thinking about a crow as a feature-constructed type, I imagine a crow strutting in a pasture or chasing a hawk away from its nest—I now will have filled out my minimal schema a bit and, more important, added to it a spatial and temporal context for it to move about in. Even if I visualize a totally motionless crow roosting in a treetop, the imagined figure of that crow will now be contextualized in a ground—i.e., the tree and the sky beyond it—which I will need to visualize also by simulating, albeit preattentively, a series of deliberate saccadic shifts in my mental gaze (Kosslyn, 1980; Kosslyn et al., 2006). A dream is similarly composed of a series of scenes, often quite disconnected, because the dreaming brain is much less able to explore its episodic imagery than is the waking brain. In subsequent chapters I will discuss some further implications of episodic memory, including its crucial relation to narrative.

When we turn to consider the principal functions of inner *action* in the narrow attention mode, we encounter simulations of action performed by, and limited to, the motor circuits of the brain. Among these simulations are those generated by the mirror neuron system. Since the common primate ancestor of humans and macaques (ca. 20 mya) was presumably equipped with these neurons, it has been proposed than an evolutionary continuum exists between mirroring and imitation, from inner simulations of action to outer enactments, e.g., the transmission and practice of fire making and tool making. The hypothesis that the sequential seriality necessary to make and use tools prepared the brain to produce and interpret semiotic sequences, such as sentences, must now be reformulated. It now seems that the mirror neuron system, a feature of the primate brain that emerged over 20 mya, provided the common scaffolding for all sequential, imitative behaviors, from fire making and tool making all the way to language.

The mirror neuron system predisposes us to imitate the actions we perceive, but we can also visualize ourselves in action by drawing on procedural memory. When we choose to simulate motor activities, we rely on the connectedness of the visual and the kinematic imagination. In this way, we are able to rehearse a sequence of actions without actually moving our limbs. Vittorio Gallese (2008) has called this "embodied simulation," or in the more felicitous Italian, "simulazione incarnata."

As I have tried to show in this chapter, consciousness, both in representation and in action, reveals a two-fold pattern that is at once broad and narrow. Its broadness consists in its capacity to perceive multiple items of information, perform multiple motor responses to those perceptions, and accomplish each of these perceptual and enactive processes at the same time, i.e., in parallel. Its narrowness consists in its capacity to focus its perception upon single items and to respond to these one at a time, i.e., serially. It can switch from one to the other—zoom out to take in a broad field of data, then zoom back in to a series of minute particulars. But this alternation is not like the all-or-nothing flicking of a toggle switch. Alert consciousness depends on a continuously variable proportion of broad and narrow attention, a duality that in practice constitutes a dyad. A healthy visual system, for example, must combine peripheral and central vision, whereas the permanent loss of one or the other creates serious disability: tunnel vision, the loss of the broad peripheral field, and macular degeneration, the loss of the narrowly focused central field.

Dual-Process Theory, which I outlined at the start of this chapter, arranges human cognitive skills in two columns of paired opposites, S1, shared with other primates, and S2, uniquely human. This does not imply an evolutionary shift from S1 to S2, but rather the addition of S2 to S1. While the new skill set, over a span of 2 million years, gradually became dominant, the old set remained and continued to play a vital, indispensable role. The result has been a division of labor, which I have represented as a dyadic pattern based on the distinction between broad awareness and narrow attention, an accommodation that we find clearly replicated in the difference between the peripheral and central optical fields.

All the major behavioral changes in the human genotype that have distanced us from other animals may be associated with this steady improvement in narrowly focused attention, accompanied by the mastery of sequential actions. Narrowly focused, sequential actions also entailed planning and led to a sense of time as a uniquely human domain, a narrow pathway through a broadly extended space. All the while, that older relation to the world, broad awareness, continued to invite humans to share this world with the animals they lived among and relied on. Animals ourselves, we would never have survived if a narrow seriality had condemned us to a myopic, tunnel-like form of consciousness.

The dyadic pattern might best be regarded as an extremely flexible template, an evolutionary strategy that allowed early humans to ease their way into early Paleolithic culture without losing the mother wit of their primate past, to evolve finer and finer micrometric muscular controls and focus on more and more intricate sequential routines, while at the same time monitoring their environment and parallel-processing its multimodal streams of information.

In the chapters that follow, we will observe how this broad-and-narrow dyadic pattern underlies a number of essential human cognitive skills, ranging all the way from play behavior, imitation, and tool use (chap. 3); to visual perception, spatial mapping, and imagination (chap. 4); to intentional signs (chap. 5); to language (chap. 6); and, finally, to verbal artifacts (chap. 7).

three

Play and Instrumentality

Our search for the origins of the verbal imagination, this "feeling back along the ancient lines of advance," must take us down some secret corridors and, lower still, to chambers buried in the subsoil and bedrock of our inner landscape. What Hermann Ebbinghaus (1908) said of psychology—that it has a long past, but only a short history—is no less true of literature. The long past that still haunts this art form perhaps explains why many over the centuries have revered it as an uncanny, even sacred, instrument of knowledge. This power, I suggest, derives from the fact that imaginative literature relies on and reactivates the deepest layers of our cognitive architecture and, in so doing, shows us the way down those neural pathways of our ancestral past.

But before we ask how this art form began, we first need to ask how language began, and, before we can answer that question, we need to ask what cognitive engineering had to have been already in place to prepare humans to share their thoughts with one another. In the last chapter, when discussing the Dual-Process Theory of the mind, I pointed out that the fundamental duality was that of perception and action and that this, as well as the other dualities, e.g., figure–ground and serial–parallel, were complementary opposites. I went on to propose that these paired dualities differ from one another in the same way: one is represented as a broad field, the other as a narrow field, the two forming an underlying two-part pattern that I called a dyad.

There are, of course, several other ways to explore answers to the question of language origins. We can examine the behavior and neuroanatomy of our closest primate cousins. We can observe the way infants develop speech. We can study how certain brain lesions affect the speech of adults. We can, thanks to recent advances in brain imaging, monitor how the intact brain performs verbal tasks. We can also analyze language itself and, through a process of reverse engineering, infer the preadaptive mechanisms upon which it was built. However we approach this question, we need to avoid a teleological bias: just because we may believe that language and its intricately wrought artifacts are the crowning achievement of some 6 million years of evolution, it does not follow that, at any point prior to the first community of speakers, language was an inevitable, much less an intended, outcome. The points through which our hominid ancestors passed on their way to language may be enumerated as a linear series, but each adaptive point along the way can merely represent one of several branched paths. Our species' arrival at every point must have been the result of a mix of environmental changes, genetic predispositions, random mutations, and successful choices—a long, tortuous journey directed continuously by the pressures of natural selection.

Donald's Four Stages of Consciousness

As I try to feel my way back along this ancient line to a sufficiently early point in (pre)human evolution, I will adopt the broad chronological model provided by Merlin Donald in his books *Origins of the Modern Mind* (1991) and *A Mind So Rare* (2001). I have already outlined his four-stage sequence, but now I will fill it out with a little more detail. Then, focusing on his first two stages, I will propose two behaviors that seem to me to require closer examination as traits preadaptive to the emergence of language and its verbal artifacts.

But first, a few clarifications. By "consciousness" Donald simply means a state of nonsleeping mental activity. Most mammals have gradations of consciousness—"active, vigilant, and wide-awake states and those that are passive, unfocused, and marked by a reduced level of activity" (Donald, 2001:118). How humans came to possess the neural architecture that characterizes their specific style of alertness is the story he has endeavored to tell over the past two decades, a four-stage story, each stage marking individuals' closer and more productive engagement with their environment. Over the past 2 million years, it was the cultural environment that increasingly drove evolution, for "cultures are more efficient than individuals at exploiting the fitness value of genetic variations, which might otherwise have a negligible impact" (Donald, 2001:259).[1]

We must also be very clear about what Donald means when he uses the word "stage." It is a period marked by the onset of a newly emerging adaptation that, building upon previously expressed traits, confers reproductive advantages on its possessors, advantages that promote population growth, which in turn helps spread this adaptation. We must keep in mind that this process in which an advantageous rewiring of the neural network selects some hominids to be our ancestors and other lines to go extinct can take vast periods of time to complete. His first stage took 20 million years, his second 2 million, his third less than 250,000, and his last a mere 5,000. The decreasing time scale of each stage suggests that a kind of Moore's Law has governed the evolution of that original "wet computer," the human brain.

We must understand, moreover, that a "stage" is not a "phase," in the sense of a temporary condition. The evolutionary account that Donald narrates does not feature stages that a species "passes through," as we commonly say of children who are exhibiting predictable age-determined behavior. When one evolutionary stage and the neural structures that characterize it are superseded by another stage, the older structures are never wholly eliminated, but rather serve as a platform upon which the next structures are built. Much of each older stage is either still in use or remains in reserve, available, if needed, to be retrofitted for later uses. (Cf. S1 in relation to S2 in Dual-Process Theory.) As Donald (2001) tells us, these stages are "successive layers in the evolution of human cognition and culture. Each stage continues to occupy its cultural niche today, so that fully modern societies have all four stages *simultaneously present*" (260, italics added).

Here again is an outline of his four stages:

1. **The Episodic.** Associated with the perception and storage of whole events, this stage evolved with the primate apes.
2. **The Mimetic.** Associated with the communication of thoughts through actions (e.g., gestural representation and the teaching of skills through showing), this began among early hominids and became fully developed in *Homo erectus*.
3. **The Mythic.** Associated with linguistic communication (telling, as distinct from showing, i.e., symbolic signs, as distinct from indices and icons), this may have appeared as early as the beginning of our species, *Homo sapiens* (ca. 500,000 years ago) or, somewhat later (ca. 200,000

years ago), when the first anatomically modern humans appeared, the subspecies *Homo sapiens sapiens*.
4. **The Theoretic.** Associated with the inscription and external storage of symbolic signs—writing—some 5,000 years ago, this led to the development of literate cultures and what we customarily refer to as "civilizations."

Now, for a little more detail. In the episodic stage, first fully realized in primate apes, the brain achieved the capacity to integrate hundreds of separate percepts, "batched together in coherent chunks" (Donald, 2001:201). An animal able to take in and organize a wide array of information into a single "event perception" of that size no longer relies solely on instinct and conditioned reflexes but can now assess and adapt itself to novel situations. The social intelligence that episodic cognition enhanced, and also benefited from, comprised "theory of mind," the understanding that other conspecifics have conscious thoughts, as well as "mind-reading," the ability to interpret their intentions to some extent. Determining the truthfulness or deceptiveness of others' responses to one's emotive and gestural signs, comparing the details of this ongoing episode with similar episodes stored in long-term memory, and then modifying one's own responses accordingly—all these capabilities took time to evolve and eventually required increased brain capacity.

In modern humans such episodes can extend in duration well beyond the interval we identify with "working memory"—seven items, plus or minus two, within a period of 20 seconds (Miller, 1956). According to Donald (2001), a cognitive episode is usually much longer than this span, though shorter than intervals storable in long-term memory. He therefore suggests we regard the episode as a period of active cognition, an event contained within an "intermediate time frame" (2001:47, 195–200), the duration of which in modern humans can sometimes extend beyond an hour (Ericsson and Kintsch, 1995; Donald, 2007b). An example of this, he suggests, is an animated conversation among several persons. "Such events are stored as single, unified episodes, and future behavior is affected by memories of such episodes. This memory for specific, coherent, detailed events is the essence of episodic cognition" (Donald, 2001:201).

Donald's second stage, the mimetic, represents the further socialization of our early ancestors by supplementing mind-readable involuntary indices with deliberately communicative gestures. "Mimesis is an analog style of communication that employs the whole body as an expressive device. Mimesis is really about acting. It manifests itself in pantomime, imitation, gesturing, sharing attention, ritualized behaviors, and many games. It is also the basis of skilled rehearsal, in which a previous act is mimed, over and over, to improve it" (2001:240). Such repetition served "as a mode of cultural expression and solidified a group mentality, creating a cultural style that we can still be recognized as typically human" (261). As we know, a structured mimetic performance, such as a ritual or an athletic event, can enlarge that "intermediate term time frame" to an episodic unit of many hours (Donald, 2007b).

Mimesis is also the basis of teaching and learning. Techniques of hunting, food gathering and preparation, fire making, and shelter construction—all these had to have been transmitted through visual observation and motor imitation. The young mimed their elders; the less skilled, their more skilled neighbors. Once a mimetic performance was stored in episodic memory, it could be repeatedly practiced and, once it became a mimetic skill, could be stored in procedural, or motor, memory. The sensory and motor intelligence of prelinguistic humans had now prepared them for their next evolutionary stage, one in which that intelligence could be communicated through a nonanalog channel—speech.

An analog mind, which up till then had been the sole cognitive operating system, receives impressions and rearranges its neural networks accordingly. These networks "form impressions in essentially this same way a time-exposed astronomical photograph does, by passively gathering data over time. Several exposures to an object allow neural networks to extract consistencies in the world that relate to the object.... In effect, the neural net classifies the world, without preconceptions about what the major classes ought to be. In contrast, symbolic computation takes in the world in prepackaged categories. It is given the major classes of experience in advance" (Donald, 2001:281).[2]

Once language evolved, the episodic bonding that mimetic culture had enhanced was marked by a further enlarged capacity of the human brain to input and manage ever longer and more complex interpersonal events. Self and other, even when they do share objects of attention, find they can have different motives for doing so, different feelings, differently remembered experience associated with their current objects of attention. Mind reading requires each to consider these differences even in the process of cooperation and to do so on multiple, simultaneous levels. Language not only communicates thoughts—it also layers, embeds, and imbricates them, so that now, "under the right circumstances, we can maintain several parallel lines of thought, each in a different mode.... Running frames within frames concurrently is routine for our species.... Our human cognitive style is linked to this multifocal consciousness, and language, in particular, is highly dependent on this feature" (2001:258–59).

What Merlin Donald describes here is dialogic give and take, the sort of improvised oral discourse that challenges our episodic-cognitive skills. But by naming this preliterate speech stage the *mythic* stage, he chooses a word that is generally associated with narrative compositions, rather than free-form conversations. "Mythic," it seems to me, more appropriately highlights the oral/literate divide, the distinction between narrative format, with its memorial storage, and written information. In preliterate society, the distinction between (1) improvised oral discourse and (2) a traditional oral composition—myth—is extremely important. As I proposed in chapter 1, improvised oral discourse forms the basis of rhetoric, whereas oral composition (i.e., the making of preliterate verbal artifacts) forms the basis of poetics.

Finally, when he comes in his story to the invention of writing, he is positioned to confirm his central premise: the primary trait that has made us different from all other species has all along been our ability to draw upon information external to the individual brain. The evolutionary stages that demonstrated how consciousness could be shared (the episodic stage), reenacted (the mimetic stage), encoded and communicated orally (the mythic stage), and then inscribed and made textually available to others are each milestones that, however twisting the journey, mark a steady trajectory from egoistic isolation to social integration. Reading this "big history," we, the sharers of the final theoretic stage, can recognize our ancestors as the men and women who opted for mutual trust, in the face of considerable risks—deception, information management, and thought control. We, the beneficiaries of this 2.5-million-year-long leap of faith, we with our bookshelves and our high-speed connections to the Internet, are the children of risk-taking parents. Moreover, as Donald reminds us, we deeply embody in our brains that ancestral heritage, so deeply that our present includes their past.

Play in the Episodic Stage

As we find ourselves engaged in a particular event or activity at a particular place and for a particular length of time, we experience it as an ongoing episode. This was how Merlin Donald used that term in speaking of the increased capacity of the primate brain to process information in working memory and thereby extend the limits of the durative present. Any episode, be it experienced in the here and now or retrieved from long-term memory, is by definition a framed activity. For example, I start my walk, I have my walk, then I end my walk. This same episodic structure corresponds to the three-part division of the narrative plot (*muthos*), its "beginning, middle, and end," as Aristotle says in his *Poetics*. Whatever I happen to do before I start and after I end frames this activity, which, while I am doing it, constitutes a particular routine, or "script."

For animals, human and nonhuman alike, play has always constituted one recognizable type of extended episode, distinct from the everyday routines of biological maintenance and survival. Why most birds and mammals exhibit play behavior of one sort or another has perplexed philosophers and led ethologists to propose a number of explanations: play trains the young to respond to real-life threats and opportunities, safely discharges excess energy, establishes dominance ranking, stimulates inventiveness, and hones sensorimotor skills through the manipulation of objects. Too often, however, play is defined not by what it does but by what it fails to do: play is generalized as any activity that does not immediately serve some practical purpose. To make matters even worse, play-like activities, such as object exploration, object manipulation, and tool use often overlap and lead to hazy definitions (Ramsey and McGrew, 2004:90).

To serve my own purpose, which is now to locate traits preadaptive to language, I will consider play as a framed, scripted episode within which particular sign values are transformed. This will entail examining the three main ludic categories—object play, social play, and imitative (or fantasy) play—in terms of *pretense*, a form of magical thinking that assumes a specific attitude toward this episodic activity. These three categories appear in most of the literature, but I take them as a triad from David Lancy (1980) where he calls them the three "play-complexes." The use of pretense as the defining feature of play I take from Alan Leslie (1987, 2002).

A play behavior typical of carnivores is solitary object play, an engagement on the part of a single animal with an inert object as though it were alive. Dogs and cats provide us with our readiest examples: a dog attacking a shoe or stick and shaking it as though it had a neck that could be broken; a cat pouncing on a wad of paper or propelling a rolled-up sock across the floor. Rather than begin by asking why they do this, we might logically begin by asking *how*, while doing so, they manage to combine two contrary thoughts—e.g., (1) this is a stick and (2) this is a rabbit. Granted, finding out *how* may be as difficult as finding out *why*, and both lines of inquiry run the risk of anthropomorphism, yet, I submit, there are some assumptions we can justifiably make. For one thing, these objects do not smell, sound, or resemble the prey they appear to represent. Furthermore, the players cannot really believe these objects are alive, for they almost never attempt to eat them. Yet what for us would be like sitting down to a heaping plate of supermarket circulars, for them seems a wholly satisfying action, an end in itself achieved by a magic trick of the mind, a paradox performed by a pretense. A play object, evidently, is *not* and, at the same time, *is* what it represents to the player. It seems, therefore, necessary to acknowledge that nonhominids engage in pretend play at the level of objects.[3]

The play of young humans reminds us that our own species continues to invest objects with a magical doubleness. As the British psychoanalyst Donald Woods Winnicott (1982/2005) observed, infants between four and twelve months begin to fixate on objects such as blankets and dolls. These, he interpreted, occupy an "intermediate area" between a child's inner psychical world and the outer world of parents and other caregivers. As such, these objects are considered partly animate, partly inert, partly incorporated within the child and partly merely "there." Winnicott called them "transitional objects" because they are used to allow for a successful transition from the autoerotic stage, in which the mother is regarded as an extension of the self, to the next stage, in which the child grants her the status of an independent other. During this transition, language comes to assume more and more the intermediate position between the child and the adult world.

Alan Leslie in his influential paper "Pretense and Representation: The Origins of 'Theory of Mind'" (1987) asked why it was possible that children, beginning at 18 to 24 months, could spend so much effort creating elaborate fantasies in which shells become cups, bananas telephones, chairs railway cars, etc., and do so at the very time they are learning the actual functions of objects and meanings of words. Why don't they become disoriented? His answer was that in their minds they are able to decouple a real (or primary) representation from a pretend (or secondary) representation, termed a "metarepresentation." Children are then able to place the latter in a kind of quarantine so that it cannot infect the semantic real-world knowledge they are now rapidly amassing.[4]

None of us entirely outgrows this need for object play. Winnicott's work with adults persuaded him that "the task of reality acceptance is never complete, . . . no human being is free from the strain of relating inner and outer reality. . . ." As we grow up, we find "relief from this strain . . . provided by an intermediate area of experience that is not challenged (arts, religion, etc.). This intermediate area is in direct continuity with the play area of the small child 'lost' in play" (1982/2005:8). Culturally coded artifacts—paintings, cathedrals, symphonies, sacraments, dramas, poems—all these, he suggests, exist to supply an indispensable play space between the adult psyche and a world that may in fact be indifferent to its existence.

Another sort of scripted episode that our prehuman primate ancestors also understood was social play. Even our earlier mammalian forebears had to have had a mutual awareness that some aggressive displays might *not* be what they seemed, just enough skill in mind reading to interpret outwardly aggressive behavior, under certain circumstances, as nonaggressive in intent. We assume so because their direct descendants, ourselves included, still exhibit this awareness. "How do animals read play intention in any conspecific? Cooperative social play may involve rapid exchange of information on intentions, desires, and beliefs" (Bekoff and Byers, 1998:xvi). Gregory Bateson (1972:177–93) had his own term for this pre-play exchange: he called it a "metacommunication." (In Leslie's lexicon, a metacommunication would be an invitation to form a "metarepresentation.") This was a signal that asked the addressee to enter into what Bateson called a "play frame" (cf. Leslie's "quarantine") within which actions (e.g., chasing, nipping, and sparring) would not indicate the sort of hostility that might otherwise lead to injury or death. Bateson used the example of a kitten playing with its mother: kittens and cats must somehow be able to frame their encounters in such a way that a nip denotes a bite without provoking the sort of escalation that this association might otherwise lead to. Here again we find pretend play, albeit stereotyped and instinct driven, that is not object centered but instead behavior oriented (Collins, 1991a:xxiii–iv; McBride, 1971).

Unlike object play, which in its typical form is the engagement of an individual with an inert object as though that object were alive, social play typically involves two individuals as though they were in mutual conflict. In object play, with no metacommunicative signal necessary, it is simply the meaning of *it* that is redefined; in social play, the reciprocal intentions of *you* are redefined. Through its power to redefine behavioral indices, social play (also called "rough and tumble") introduced into mammalian cognition a special kind of framed episode in which one distinct event was embedded within another: event 1 is a fight, while event 2 is *not* a fight. In semiotic terms, the metacommunication is a preliminary *index* that negates in advance all subsequent *indices* of aggression (in event 1) and transforms them into mutually recognized *icons* of aggression (in event 2).[5] So, the indexical communication that precedes such an episode means "I invite you to enact with me a script in which what we two do will indicate aggression but not lead to injury, because we agree that it merely *looks as though* it indicates it." This formula, essential to social play, keeps the meanings of these two simultaneously performed events, each with its own sign function, separate and distinct. In social play, as in object play, the same as-if principle is invoked: something seems like something else and, by virtue of its iconic function, is known to be distinct from what it stands for.

Let's now consider some further implications of play during Donald's episodic stage. Imagine a settlement of Australopithecines. It is a sunny midday about 3 mya. Grandmother, a gray-haired, thirty-year-old elder, is resting from her work of gathering berries and watching her four-year-old granddaughter, who now crawls up toward her. Grandmother closes her eyes and begins to breathe deeply. As she does so, the girl plucks a grass stalk and is about to tickle her ear when the old female with a wide-eyed, gobble-you-up expression grabs her and gives her a vigorous shaking. Then she laughs and resumes her recumbent position. The girl, too, laughs and, a few seconds later, creeps up again with another stalk. This time grandmother has one eye open and one eye closed as she breathes deeply and sonorously. The girl sees that the woman sees her, yet stretches out the stalk anyway. The old one grunts, lunges for the girl, and they both laugh. This has become a game in which both parties pretend to be deceived. Henceforth, between them the gesture of the wink might become a sign that things are not really what they seem, that, for example, the strutting of the clan leaders should not always be taken at face value—it, too, might be a sign decouplable from the actual significance of these elders.

Such episodic social play depends on mutual mind reading, a recursive metacognitive I-know-that-you-know-that I-know in which the *that-which-is-known* is that the aggressive index is really only an icon. This vignette also illustrates one of the possible gestures that voluntary eye movements can convey—in this case, "I am asleep" and "I am awake"—a nonverbal ironic stance that can also signify "I don't know" and "I do know." At every level, this interchange between grandmother and granddaughter is frame-doubling, which needs to be carefully distinguished from frame-*shifting*, for, while both procedures decouple sign from meaning, frame-shifting depends on another's miscomprehension ("I-know-that-you-*don't*-know"), an episode in which only one party is at play. This shifting, as Seanna Coulson demonstrated in *Semantic Leaps* (2001), is essential to comedic play of the joke and sight gag variety. It starts by leading us to believe that one particular interpretive frame governs a given episode, then suddenly whisks that away, revealing the real frame.

In his essay, Bateson extended his play insights into the topic of fantasy. Viewed in psychoanalytical terms, fantasy is an expression of "primary process thought," which is the way the dreaming mind defines reality. "Secondary process thought," in touch with the "reality principle" can interpret dream images and waking fantasies as metaphorical by consciously enclosing them in an "as if" play frame (1972:184–91). Bateson's theory of double framing in both play and fantasy is remarkably consistent with Dual-Process Theory as I described it in chapter 2. Since physical play and mental (metaphorical) play involve double-framed cognition, they are further examples of a successful, dyadic integration of primate and uniquely human thought.

The evolutionary advantages of frame-doubling in physical play behavior include the opportunity to practice fight-and-flight motor programs in the safety of a band or family. Since social play signifies alliance, this behavior also reinforces group solidarity. But the pertinence of mammalian social play to the emergence of language lies in its preadaptive uses. First, it prepared our hominid ancestors to perceive a social encounter as an episode, framed within a familiar context. Secondly, social play exercised the mirror neuron system, those cells that convert visually perceived motions of conspecifics into motor neuron firings that mirror those movements.[6] Finally, as hominid communities grew in size, this particular social script could be extended from individuals in one-on-one play to whole groups in two-sided coordinated play.

Social play seems to have evolved later than object play. One indicator of its increased cognitive complexity is that social play can accommodate elements of object play within it. Two dogs, for example, can struggle to possess a stick or a rag as though they were fighting over a fresh kill and all the while adhere to the rules of play. Human sporting events, both the one-on-one and the team varieties, also combine object play within social play. Such rule-governed agons focus attention on an object that signifies a valuable property, an object that players on one side must keep away from those on the other side, successfully control, and stow away in some secure location. The same blend of object and social play obtains in the less strenuous games played on boards and tables, using cards and counters. As I will later propose, the evolution of communicative codes, both gestural and vocal, that culminated in language adapted mammalian play behavior as its driving force (Bruner, 1972; Siviy, 1998).

Instrumentality in the Mimetic Stage

Most evolutionists agree that tool making and language are the two traits that have most distinguished and distanced us from all our fellow animals and that tool making preceded true language by at least 2 million years. But when the discussion turns to if and how these skills may have at some point co-evolved, agreement is hard to come by. So, were humans, their hands occupied with the manufacture and use of tools, now obliged to "gesture" with vocal sounds? Was the capacity to remember and employ sequences in making things used subsequently to string sounds together in some meaningful syntax? And how did social organization affect the development of both skills? My own observations, which follow, are based on the assumption that the visuomotor and proprioceptive experiences that our Lower Paleolithic ancestors had when using simple manual tools were virtually the same ones we have today when using comparable tools. A phenomenological approach to tool use, supported by cognitive psychology, ought therefore to help us form valid inferences that may shed light on the origins of language and verbal artifacts.

It is safe to assume that hominids used found objects, such as boughs and stones, as tools for millions of years before their descendants learned to fashion these into clubs, scrapers, hand axes, and other products. The use of found objects is probably as old as the earliest great apes (20 mya), but the earliest modified (i.e., handcrafted) tools date only from about 2.5 mya, the beginning of the Lower Paleolithic era. Discovered in the Olduvai Gorge of northern Tanzania, this manufacture ushered in the period that Donald characterizes as the mimetic stage.[7]

With all our technical sophistication, we modern humans do still employ improvised tools. The ability to perceive the distinguishing details of edible, medicinal, and poisonous plants, to catch fish and animals, to improvise shelters and coverings—these survival skills and the tools we find to practice them are no mean achievements and continue as part of our hominid repertoire. Walking in hilly terrain, we might steady ourselves with a straight stick found beside our path, drink spring water from a burdock leaf curled into a cup, or hammer open a hickory nut between two stones. Such ad hoc instruments as a stick, a leaf, or a pair of stones, once used, we would probably toss by the wayside. But suppose we carried with us a knife or a fishhook and line. These we would keep, because reusable tools like these cannot be readily found where and when we need them (Napier, 1980:148–49).

Let us now for a moment imagine a scene some 1.5 mya. It is a clearing, a settlement of *Homo ergaster*. A man squats on his haunches. His son is standing beside him. The man gestures toward an eland skin beneath which lies a cache of oval stones. The boy brings him one, which the man takes in his left hand and proceeds to strike with another stone. Before each stroke, he repositions it with his thumb to slip it outward toward his fingers. As soon as a sparkling row of concave marks appears on one side, he turns it over and patiently makes the same pattern on the other side. Eventually, the shining inner core of the stone is revealed. He now begins the delicate work of chipping one face of the stone in such a way that the tortoise-like surface has a raised spine down the center of the oblong. Finally, with a perfectly aimed blow at the top of the stone he splits it, detaching a razor-sharp flake the size of the boy's palm. Carefully watching, the boy is learning that there is a proper order in the transformation of a roundish black stone into a gleaming hand ax capable of slitting the hide of a mastodon.

What the boy was learning and was soon to practice was a technique that did not depend on the season or the circumstances of his family or clan. Stored in procedural memory, this technique consisted in autocuable kinesic schemas, i.e., motor routines of eyes, arm, and hand that could be voluntarily initiated (Donald, 1999). Context-free, this learned behavior could be reproduced whenever and wherever it was needed. Unlike episodic memory, which stores the specific but ignores the general, procedural memory stores the general and ignores the specific. Unlike semantic memory, which is general knowledge in a private archive, this is socially stored knowledge transmitted by demonstration and imitation to one or several persons, who may later demonstrate it to several others. Each of these may teach it to several more, thus swiftly spreading it in a radiating network. Though it would not be shared with unfriendly others, friendly contact with other groups would ensure that a useful technique would replicate swiftly.[8]

Tool making, the crowning achievement of the mimetic stage, established the advantages of slow, patient, repetitive actions. The careful flint knapping that produced the tapered spearhead, the scraping that produced the long shaft, and the months of instruction and rehearsal of the throw and the thrust—all these mimetic behaviors paid off in the successful hunt and the sharing of meat with family and clan. With artifacts such as these, humans were no longer dependent on what they found lying about, the "naturefacts" that served as improvised tools. They now had portable equipment they could reuse and, when necessary, repair or improve. Products of a culture that had learned the value of repetition, these portable tools prepared their owners to venture into unfamiliar territory, protected them if they encountered danger, and, as valuable property, could be used as gifts or items of trade.

There is also something magically metamorphic about both the found tools and made tools. In *Being and Time* (1962/1927), Martin Heidegger noted how, from our human perspective, a thing can shift its status suddenly from that of an object to that of an instrument. Situated "out there," somewhere, anywhere, in the world, all things may be categorized as either "present-to-hand" (*vorhanden*), i.e., there simply as objects, or "ready-to-hand" (*zuhanden*), i.e., usable equipment. If, for example, my path in the woods leads me to a boggy spot and I see a half-rotted log nearby, I might decide to lay it down across that spot and use it as a kind of bridge. When I do so, I convert that object into an instrument, something *vorhanden* into something *zuhanden*. Then, having crossed, should I glance back at that log from farther up the path, I find that it is no longer a bridge-like instrument but has turned itself back into an object. Of course, it could have already turned itself back into that object if, as I stepped onto it, it had broken apart (Heidegger, 1927/1962:92–93, 96–107).

A similar shift could befall a made tool. Consider a hammer lodged neatly in a toolbox beside a set of screwdrivers and pliers of different designs and graduated sizes. Nestled there, it is an object among other objects, a testimony perhaps to the owner's technical finesse or to his obsessive orderliness. But if I need it and he lends it to me, the hammer begins to change. I heft it, sense its balance and comfortable grip, and later swing it to nail a piece of clapboard to a wall stud. Doing so, I feel how "ready-to-hand" it is—in fact, I sense it as an extension of my hand. Alternatively, if, after a few blows, its head flies loose, this instrumental extension of my hand becomes an object once again. As Heidegger remarks (whose hammer example is always ready-to-hand, by the way), this piece of equipment becomes at this point conspicuously unready-to-hand (Heidegger, 1927/1962:102–3).

Michael Polanyi pondered further on what it might mean for such a tool to be an extension of its user:

When we use a hammer to drive in a nail, we attend to both nail and hammer, *but in a different way.* We *watch* the effect of our strokes on the nail and try to wield the hammer so as to hit the nail most effectively. When we bring down the hammer we do not feel that its handle has struck our palm but that its head has struck the nail. Yet in a sense we are certainly alert to the feelings in our palm and the fingers that hold the hammer. They guide us in handling it effectively, and the degree of attention that we give to the nail is given to the same extent but in a different way to these feelings. The difference may be stated by saying that the latter are not, like the nail, *objects* of our attention, but *instruments* of it [my italics]. They are not watched in themselves; we watch something else while keeping intensely aware of them. I have a *subsidiary awareness* of the feeling in the palm of my hand which is merged into my *focal awareness* of my driving in the nail. (1958:57)

In this example of tool use we are co-conscious of both nail and hammer, but as Polanyi realizes, we attend to each in a different way. The object of our action receives "focal attention," while our instrument receives "subsidiary attention," a distinction that precisely corresponds to what I have proposed as the dyadic pattern. To use this tool successfully, we must coordinate these two functions: we must focus narrow attention on the nail, while we remain broadly aware of the feel and heft of the hammer as we swing it toward its object. But is the hammer the *object* of broad awareness (our subsidiary attention)? No, Polanyi calls it an *instrument,* not an object. Then he muses: "Think how a blind man feels his way by the use of a stick, which involves transposing the shocks transmitted to his hand and the muscles holding the stick into an awareness of the things touched at the point of the stick" (55–56).

Over a decade earlier, another philosopher, Maurice Merleau-Ponty (1945/1962), had indeed thought of how a blind man feels his way and concluded:

The blind man's stick has ceased to be an *object* for him, and is no longer perceived for itself; its point has become an area of sensitivity, extending the scope and active radius of touch. . . . In the exploration of things, the length of the stick does not enter expressly as a middle term. There is no question here of any quick estimate or any comparison between the objective length of the stick and the objective distance of the goal to be reached. To get used to a hat, a car or a stick is to be transplanted into them, or conversely to incorporate them into the bulk of our own body. Habit expresses our power of dilating our being in the world, or changing our existence by appropriating fresh *instruments*. (143, italics added)

When we incorporate them, tools give us the power to extend ourselves outward into the world, and, as we extend, they magically vanish into our foreground. Yet, as Polanyi pointed out, this vanishing is not absolute: the tool continues to be held in subsidiary awareness, that particular level of alertness we must continue "intensely" to maintain.

This phenomenon to which all three philosophers refer is a doubleness fundamental to all sensory modalities and is akin to figure–ground perception. The object, when it becomes *zuhanden*, becomes an instrument and as such enters the ground of its user, while its point of contact with the world becomes that tool user's *vorhanden* object. To use the hammer-nail example, the hammer, joined to the hand-arm-shoulder of the worker, magically vanishes into that worker's ground, and the nail becomes the focalized figure.[9]

We tend to consider technologically simple communities around the world as prone to "magical thinking" and assume the same of our Paleolithic ancestors. Perhaps so, but if we define magic as this power of things to metamorphose, magic is perfectly natural, and cognitive evolution, one might argue, accelerates each time a new magic trick is learned. We have been considering tool use as a sort of magical extension of the user into space, a means of effecting changes at a distance through a sequence of skillful motor adjustments. In this respect and in others, which I will suggest in later chapters, tool use is a precursor of other magic—that of speech and of verbal artifacts.

Imitative Play in the Mimetic Stage

Play, as I have considered it, involves pretense and transformation, the pretense that indexical signs can transform themselves into iconic signs. In other words, play happens when the mind decouples things and actions from their normal real-world contexts and makes them represent other things and actions. In object play, as we have seen, a player pretends, while knowing otherwise, that an object has been transformed into something else—e.g., a living thing, prey, or some other item of interest. In social play, each player pretends that the other has been transformed into an enemy. Now, in imitative play, players pretend that they have been transformed into other persons. (Cf. the idiom "The actors *play* different characters.")

Along with tool making, the mimetic stage introduced this latter form of play, a specific subset of mimesis that was to lend its own magical thinking to the evolving structures of consciousness. As imitation, this form of play is not merely a repeated, coordinated action, or the sort of mirroring observed in social play, nor is it the repeated demonstration and rehearsal of a skill or the conveying of information through iconic gesturing. It is all these and more: it is deliberate role playing, the representation, on the part of one or more persons, of one or more *other* persons, animals, or objects. In this respect, too, it is different from social play, which typically involves two identified individuals engaged in enacting a mutually recognized play script—e.g., two dogs barking and taking turns chasing one another, or two humans exchanging soft punches to their shoulders to communicate good-natured bonhomie. If, for example, I give my companion a friendly shove, I am, after all, not imitating someone else. It is *my* act. *I* intend it. In short, while social play is initiated by a person whose understood intentions contradict his apparent actions, imitative play is initiated by a person who imitates someone or something other than his own understood identity.

Imitative play differs from social play in yet another respect: it allows for multiple participants (actors) and the presence of a spectator/audience. As spectacle, this play helps establish a sense of social solidarity by arousing and synchronizing the emotional responses of the community through the presentation of an episode, or a series of episodes, that has inherent relevance to that audience—typically an action with life-and-death implications. During such high-arousal episodes, not only is the time duration for all in attendance extended into the "intermediate term," but their moment-to-moment input flow is quantitatively enlarged. The movements of the performers will therefore evoke incipient motor responses in the mirror neuron systems of their viewers, simulations that will prompt them to empathize with, and thereby more accurately interpret (mind read), the dramatis personae.

All those present during this mimetic act understand it as play: the imitator and the imitated, the human signifier and the meaning signified, are at once coupled and decoupled. In respect to the dyadic pattern, the imitated (the fictive persona) is narrowly focused on, while the fact that this is an imitator (an actor) is held in background awareness. To consider it simply as the action of a mimer or as the interpretation of spectators risks ignoring the fact that, like social play, imitative play is an intentional communicative performance that must be understood as such. If I seek to amuse my guests by imitating the walk and speaking style of a public figure and they think I have injured my leg or suffered a momentary slurring of speech, the imitative play has been incomplete, which is to say: it did not happen.

As in social play, a play frame circumscribes an imitative play performance, and, whether the latter is a ritual or a heroic pageant or a piece of informal clowning, the player first needs to metacommunicate the *index* that this is a performance, not a "real" action. Only after that is indicated can the performer present indexical actions (movements, gestures, vocalizations, etc.) as *iconic* of some other person, action, and condition. Imagine, for example, the following scenario (since I have been assigning dates, let's make it 200,000 years ago): A person appears before the assembled clan. He is a well-known elder with a particular personality and life history, but now he is imitating a mighty hunter tracking a bear. *That* he is imitating is conveyed by a metacommunicative index, perhaps the fact that this takes place in an area marked off for such performances. *What* he is imitating through his own indexical actions—his facial expressions, his proud gestures, his tensely alert gait—is the icon of a person that all know he could never have been, even in his youth. This is an intentional icon, a conscious simulation that audience and actor agree to engage in. If he uses objects (theatrical props, masks, etc.), his indexical behavior toward them confers iconic powers on them too, so that they become that which his actions seem to indicate they are. Coming together with family and neighbors to view this performance is itself an episode—a real event—that within the limits of a particular place and time frames a nonreal event. Narrativized double-framed imitative play (i.e., icon framing index and index framing action), thus appears to be the prototype of all later forms of fictionality.

Imitative play, the last play complex to evolve, is a special human achievement. Though some cases have been reported among great apes, as a general rule only a human has the ability to step back and view him- or herself from the vantage point of another, an insight, by the way, that also permits us to deceive one another by imitating someone we are not. An ape can deceive through devious actions (indexical fraud), but a human can also deceive by passing himself off as another person (an iconic fraud). It is only when this impersonation is meant *not* to deceive that it becomes imitative play. Writing toward the end of the nineteenth century, the French psychologist Victor Egger (1881:130) commented:

> [I]n every instance of play, in every make-believe, the soul divides and the earnest actor masks a skeptical spectator. . . . [I]n play, generally, the individual ego affirms and denies itself simultaneously or at imperceptible intervals. . . . In so doing, the mind does not believe it is contradicting itself: out of this affirmation and this denial it makes a synthesis and this synthesis is the very essence of play and drama.[10]

In play, Egger proposes, affirmation and denial are not contradictory states of mind. Instead, the two opposites are synthesized and thereby achieve the status of a dyad. This doubleness is also characteristic of visual imagination, which is indeed a form of mental play. A mental image is consciously recognized as the simulation of a percept, but it is never confused with that percept, unless the imaginer is delusional.

Preludes to Language

This chapter has been a sort of genesis story intended to sketch out some of the major cognitive advances that must have preceded the emergence of language and verbal artifacts. Why one genus of primates could outdistance its related genera and why, within it, one species, *Homo sapiens*, could outlive all its related species might best be understood in terms of its capacity to extend consciousness into time and space. As Merlin Donald argued, our primate ancestors had extended their awareness of the moment by expanding working memory to encompass perceptual events of more and more complexity. Gifted with a more capacious episodic consciousness, their social encounters introduced them to one another as potentially helpful allies in their competition with other predators and scavengers. As a consequence of their enhanced social intelligence and their knack for bipedal locomotion, they were able to survive, find a niche, and expand their African habitat.

This evolutionary period (20–2.5 mya), which Donald has called the episodic stage, was marked by a slow, but steady, increase in brain size, perhaps in response to a need to manage a broader network of social relationships and a more diverse geography. The Australopithecine brain was not much larger than that of a modern chimpanzee, but with the next stage, the mimetic, the brain had doubled its size. The human genus inherited this older, genetically hard-wired architecture and the social intelligence that had already evolved could now provide hominids with a repository of ecological, or object, information. In short, with the mimetic stage came socially distributed cognition and material culture.

These evolutionary stages, Donald reminds us, are cumulative. As Steven Mithen (1996) succinctly puts it: "[T]he architectural plans may have been continually tinkered with, but no plan ever started again from scratch. Evolution does not have the option of returning to the drawing board and beginning anew; it can only ever modify what has gone before. That is, of course, why we can only understand the modern mind by understanding the prehistory of the mind" (65). If we say that the mimetic stage inaugurated *cultural* evolution through a meme-driven (as distinct from a gene-driven) series of modifications, we simply mean that the biological mechanisms already in place made cultural innovation possible and thereby set in motion a new, accelerated process of change. Michael Tomasello has called this process

> a kind of cultural ratchet, as each version of the practice [of any skillful behavior] stays solidly in the group's repertoire until someone comes up with something newer and more improved. This means that just as individual humans biologically inherit genes that have been acquired in the past, they also culturally inherit artifacts and behavioral practices that represent something like the collective wisdom of their forebears. To date, no animal species other than humans has been observed to have cultural behaviors that accumulate modifications and so ratchet up in complexity over time. (2009:xi)

The oldest extant indicators of cultural behavior are, of course, the stone artifacts associated with Paleolithic hunting and butchering. There undoubtedly existed other instruments made of less durable materials and used for other purposes, but stone hand axes and, later, spear heads tell us two things: humans had mastered the technical skill to reproduce these artifacts and high-energy animal protein had become an increasing portion of their diet. While Australopithecines had employed only found objects, their human descendants, by handcrafting instruments, became able to hunt larger mammals and nourish larger populations. The preservation of these techniques implies their cultural transmission from generation to generation—i.e., a capacity to imitate the actions of others and store these routines in motor (procedural) memory. Here the mimetic stage clearly built on the advances of the episodic. By requiring advanced planning and patient adherence to certain learned sequences, tool *making* extended episodic consciousness into the temporal future; by requiring deft hand–eye coordination, precise timing, and accurate aim, tool *use* extended episodic consciousness outward into the spatial here and now. Armed with such self-extending instruments, Homo erectus and related species ventured beyond the African continent and, by 1.8 mya, had begun a migration that would take them into most of the Eastern Hemisphere.

Improved hunting techniques, as distinct from scavenging and gathering, provided early humans with leisure time, and, as with most mammals, leisure leads to play. This behavior, widely observed among birds and mammals, is far older than the episodic stage. Object and social play constituted a distinct kind of episode that our hominid forebears must have found enjoyable—and *beneficial*, for the fact that their descendants have inherited this trait suggests that natural selection strongly favored the playful hominid at the expense of the dour hominid. With the onset of the mimetic stage, our early human ancestors continued these episodic play behaviors, but they must have added that new form, imitative play. This mimetic adaptation was, I suggest, an innovation no less important to the future of genus *Homo* than the invention of stone instruments had been to its past. Imitative play, expressed as pantomime, masking, and dancing, introduced a new category of reusable instruments fabricated out of repeatable actions: behavioral routines that might rightly be termed artifacts.

The semiotic structure of the play episode that I have diagrammed in figure 3.1 holds true for each of the three "play complexes:" object play, social play, and imitative play. Because of the semiotic operations implicit in it, play behavior was especially preadaptive to language, but play behavior may have had other language-related evolutionary implications. Object play, for example, has been considered a precursor of tool use and manufacture, which in turn is believed by some to have been preadaptive to language. Social play has been associated with theory of mind, mind reading, empathetic simulation, and the mirror neuron system, which has been located adjacent to and connected with those parts of the brain (e.g., Broca's area) that control both manual gestures and vocal articulation.

In social play, we must accommodate the other, the play partner, and understand the action from that other's perspective. Thrust and parry, flight and pursuit, offense and defense—these roles must alternate. We realize here that every *you* is, from his or her own perspective, an *I*. The theory of mind thus underlies the function of second- and first-person pronouns as *shifters*. To use Alan Leslie's (1987) terms, each has to know that the *intended* content of the primary representation (the overt actions indexical of a hostile encounter) is actually the secondary representation (the friendly attitude beneath the semblance of hostility). Each must rely on an interpretation of the other's intention that justifies particular actions. In other words, social play requires the ability to assume another's perspective as well as one's own.

As Jespersen and, later, Jakobson and Benveniste used this term, a shifter is a grammatical element that can properly be understood only in the context in which it is uttered. When I say "I," it always means myself, but when you say "you," it never means *you*rself. Whoever speaks is an *I* from whose perspective every addressee is a *you*, while all beyond this speech circle is implicitly a *he, she, it,* or *they*. Just as social play involves turn taking, speech involves the give-and-take of dialogue marked by the shifting ownership of *I*. I will revisit this topic of pronoun shifters in chapter 6, "Language: Its Prelinguistic Inheritance," but, even before the emergence of language, imitative play could have destabilized the egocentric perspective. Groups that could foster an allocentric attitude among members would prove more successful at tasks requiring social cognition, e.g., hunting and child rearing. This perspective could be regularly reinforced through rituals aimed at deemphasizing individual differences, behaviors such as repetitive dancing and the use of uniform body paint, ornaments, and masks.

Figure 3.1, incorporating ideas drawn mainly from Bateson (1972) and Leslie (1987), illustrates a theory applicable to each of the three play complexes. The initial indexical signal (A) is a metacommunication, an invitation to engage in play. Once play participants enter the first play frame (B) and from that into the second play frame (C), they find themselves in a double-framed episode wherein, as Leslie put it, are "two simultaneous representations of the situation. One representation is for how the situation is actually perceived, whereas the other represents what the pretense is" (414). In my diagram, B is the broad, background awareness that frames C, which in turn frames the play action. The difference between B and C in object play depends on the player's understanding that, for the duration of the episode, the object (B) has a new meaning (C). In social play, the difference is jointly agreed upon by the two players, or teams, each of whom will exhibit an aggressive attitude (C) while preserving a background awareness (B) that the other is a mere iconic rival; in imitative play, audience and actors alike share the knowledge (B) that real players are iconically representing other persons (C).

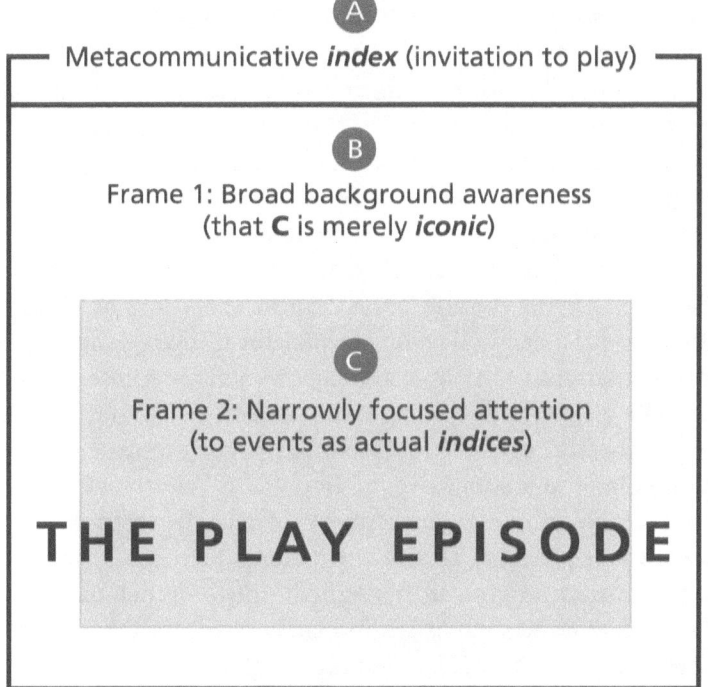

Figure 3.1 Semiotic play framing

Leslie seems to invite a semiotic analysis when he characterizes primary representation (B in my diagram) as having a "direct semantic relation with the world" (414). This perceived object or action takes its primary meaning from its place within the spatial/causal network of perceptible reality. For example (and this is Leslie's example [421]), when a mother initiates object play with a small child by picking up a banana and using it as a telephone, both mother and child recognize that the banana, as primary representation, continues to be that soft, sweet fruit with a yellow peel even while it secondarily represents a telephone. This metarepresentation happens because (1) the mother behaves indexically toward the banana as though it could function as a telephone, and (2) because it shares certain iconic features of size, weight, and shape. What play must ultimately depend on is the mind's capacity to interpret an object or activity as both indexical and iconic simultaneously in a dyadic synthesis.

For me to apply Leslie's insights to my dyadic model I must note, however, that his terms "primary" and "secondary" representation presuppose a detached, objective point of view. If we want to understand the play experience, however, we need to assume the perspective of the play participants. When we do so, Leslie's terms become inappropriate. For both mother and child, the primary, figurally foregrounded interpretation of the object is its pretend identity. He calls the latter a "secondary" metarepresentation, yet for mother and child this talking into the banana is the primary (in the sense of "focally attended to") event, while the secondary, background awareness is that this "telephone" is *really* a banana.

Though Leslie's article is exclusively concerned with children's cognitive development, its subtitle, "The Origins of 'Theory of Mind,'" leads us to consider further implications of play for human evolution and the emergence of language. As an exchange of mental representations, language entails an ability to distinguish between those that belong to an objective, real-world context (these being primary representations) and those that belong to a subjective, ad hoc, "as if" construct (these being secondary representations). The concepts implied by verbs such as *think, remember, wonder, believe,* and *dream* belong to mental states that in themselves cannot be outwardly verified. On the other hand, the concepts implied by verbs such as *see* and *know* assert, by contrast, veridical references to a perceivable, sharable reality, though, of course, they may well prove false or deceptive (422). Play behavior precedes and, as Leslie maintains, builds the child's later capacity to produce more nuanced expressions of mental states and language facilitates this development by distinguishing *is* from *seems* and both from *as if* states of mind.

It was this same behavior, I would argue, that also preceded the evolutionary emergence of language. Having prepared the mind to decouple mental states from general knowledge and transitory iconic fictions from established indexical facts, play propelled our ancestors to take the final step in that long process of semiotic decouplement and devise a system of conventional symbols only arbitrarily related to their meanings. Out of these symbolic elements, new artifacts were to be invented, crafted out of breath, heard as sound, imagined as pictures, and, thousands of years later, seen as writing. It is hardly surprising then that, born of the playful medium of language, these artifacts continue to be animated by its spirit of play.

Now, just as social play can include object play within it, imitative play can include object *and* social play and, in that respect, resembles language in its superordinate powers of representation. Finally, imitative play can be linked to mimesis and the transmission of general knowledge, two cultural behaviors that language was designed to facilitate.

To conclude: imitative play, sprouting from a seed long hidden in the episodic subsoil, finally blossomed in the mimetic stage. When it first emerged, whatever purposes it then served in a world without language—ritual actions associated perhaps with sympathetic magic, spirit worship, or commemoration of a shared past—it incorporated all the essential elements of what we have now come to recognize as verbal artifacts: collectively owned compositions that are prized, preserved, and re-performed.

four
The World as We See It

We are accustomed to think of literature, as distinct from philosophy, history, science, and the like, as "imaginative writing." By that we generally mean it represents particular persons, places, and events that an author has imagined, rather than actually experienced, or, at any rate, that readers must imagine without having to judge the truth or falsity of this pretend world.

My reason for introducing this uncontroversial notion here is first to endorse what I trust is by now another uncontroversial notion, namely, that imagination is a process that in many ways replicates visual perception. In my *Poetics of the Mind's Eye* (1991a), which was grounded largely in the work of the cognitive psychologists Allan Paivio, Roger Shepard, and Stephen Kosslyn, as well as that of the cognitive linguists Ronald Langacker and Leonard Talmy, I attempted to show how literary texts induce us to simulate actual visual events in a format that is borrowed partly from dream, partly from episodic memory. In this chapter, which incorporates more recent research on visual cognition, I hope to add to and clarify the theory of imagination I presented in 1991. Placed just prior in this evolutionary saga to the emergence of protolanguage, this chapter will introduce preliminary evidence for another claim, one that I will subsequently spell out—that visual perception and imagination formed the basis for language and, consequently, for verbal artifacts.

Language and its artifacts evolved out of visual perception and imagination: really? Well, yes . . . but not quite. When it comes to describing cognitive processes, simple linear accounts often falsify reality. As Einstein advised, we should strive to make everything as simple as possible, but not simpler. I will first follow this maxim by describing the simple structure of the eye and how that contributes to our view of the world. Then, as I explore the neural anatomy responsible for perceptual processing and action governance, I will honor the "not simpler" proviso. In this deeper anatomy we will again find a coordinated doubleness, both in the circuitry of the visual brain and in the two alternate ways we have of organizing visual space. I will go on to speculate that this dyadic pattern, which governs the way we receive and process visual data, has so profoundly impressed itself on cognitive mechanisms that it marks an outer limit of our conceptual powers.

A word of caution: the topics I deal with in this central chapter do themselves challenge our conceptual powers. At any rate, they do not lend themselves to easy exposition. Much of what I discuss is neural architecture buried deep within the brain, so in this chapter, devoted to the sense of sight, most of what I talk about lies totally hidden from sight and resistant to visualization. I do, however, revisit and build on earlier introduced themes, e.g., perception and action, figure and ground, and parallel and serial processing. In portraying how these operate, I must stretch the capacity of that old technology, words, and depend on the reader's even older brain-scanning device, the imagination.

Vision and the Visual Imagination

The eyes of a hunter-gatherer are acute. Flesh, fruit, tubers, honey hives—these are isolated, desirable objects amid a multitude of less interesting things. When those first adventurous bands of hominids, our earliest ancestors, left the safety of the forest canopy to range the grasslands of East Africa some 6 mya, they turned their binocular/stereoscopic vision toward an expanded vista. In the forest the objects they saw—and heard and smelled—had been intimately near at hand, within a hundred feet on average, but on the rolling savannahs they might make out herds and fruit trees, springs and rocky ledges, many miles away. Here was a multitudinous array of distant details that could neither be heard nor smelled but only *seen*, an environment that determined over time that only the best sighted among them would survive and transmit their optical acuity to succeeding generations.

A viewer's eyes gazing fixedly ahead survey a broad field, an oval that is nearly 160 degrees from side to side and 135 degrees[1] from top to bottom, though early hominids would have had a shorter visual field in the vertical axis due to their overarching brows. Thanks to binocular vision, there is considerable overlap of the right and left ocular fields. Each eye therefore reinforces the other in straight-ahead central vision. Since the two eyes are separated on the face, the two views they produce have slightly different spatial orientations, which the brain uses to compute the relative depth of objects, the 3-D effect. Our ancient ancestors would no doubt have spent some time, as we do, in unfocused vision, their open eyes taking in the environs with no particular object in mind, but in that perilous world they must have spent much more of their time intensely gazing about them, on the lookout for predators they needed to elude and for food they needed to capture.

On the basis of fossil and current primate evidence, evolutionary biologists assure us that the anatomy and physiology of the primate eye has remained virtually unchanged for at least the past 10 million years. At the center of the broad oval visual field just described, there appears a small, clear disk. This central field (about 15 degrees in diameter) has at its midpoint the smaller, even sharper foveal field (about 1 degree across). One way to demonstrate how these two fields work is to hold your hand at arm's length, palm inward, and try to see all its lines, creases, and contours without moving your eyes. (The experiment works equally well whether you use one or both eyes.) If you focus on the center of the palm, the mounds at the bases of the thumb and fingers seem less detailed and the thumb and fingers seem progressively fuzzier as they radiate from the palm. The central field is simply too small to encompass this space. Within it, the even smaller foveal field, densely packed with contour- and color-sensitive cone cells, is used for scrutinizing minute details. If you now turn your stretched palm outward and gaze at your little finger, only its nail (approximately 1 centimeter in diameter) is acutely visible, for it projects its image on that 1-degree disk of cells at the back of the eyes.

To recognize an object that overflows the central focal field—e.g., a person's face at a distance of 10 feet—the eyes must perform a set of rapid movements (saccades) and a set of momentary fixations. The brain then converts these part-perceptions into a complete, meaningful figure isolated from its surrounding ground. The size of this central field is necessarily small, for if the eyes could present to the optic nerves the entire visual field in perfect clarity the input would overwhelm the brain with random data. That is one good reason why, in the area outside the central field, objects appear less sharply contoured and colored.

Surrounding this 15-degree central field lies the much larger peripheral field, constructed by the motion-sensitive, but less detail-sensitive, rod cells of the retina. Yet despite its diminished acuity, the peripheral field is essential to the function of central vision, for it provides the brain with a broad, general account of the environment and of possibly important objects moving there, even in dim light and shadow. In certain circumstances, we can in fact use peripheral vision as our sole viewing mode. When, before we cross a street, we look both ways, we turn our heads only so far as to glimpse moving vehicles in our left and right peripheral fields. Consider, too, the way we search for a misplaced item—e.g., our car key, which we must have somehow dropped on the way from our parked car to our front door. If the area is too large, we do not choose to use our focal vision. That key could be anywhere, so, to save time, we shift from focal to peripheral vision by locking our eye muscles in place and slowly rotating our head, effectively suppressing the saccadic mechanism that generates sharp ocular fixations. As we broadly scan the area, we keep in our mind a generalized image of a silver-colored key and its ring. If any shape that even vaguely resembles that object registers in our peripheral field, we stop and direct our central vision upon it. In short, the peripheral field is like a broad net cast upon the world: whenever a useful object is caught in it, the spear-like central vision darts forth toward that object and transfixes it.

A common notion of folk psychology is that seeing is indeed a kind of spear throwing, but one in which the spear is made of light. According to this conceptual metaphor, known as "extramission," the eye sends forth a "ray" or "beam." Interestingly, both these words originally meant wooden projectiles, shafts, as in the metaphor "shafts of light," itself a related instance of folk physics. Accordingly, we "cast a look," "shoot a glance," have "darting eyes," "stare daggers," and so forth. Though this metaphor inaccurately represents visual perception, it does correctly portray focal vision as a conscious, voluntary turning of attention to objects.[2]

While this ancient metaphor reminds us that central vision is indeed an action, we sometimes forget that peripheral vision can also be an active process. Though it is usually a "preattentive" process, it is sometimes *intentionally* preattentive when used to perceive simple shapes and faint sources of light, as my key-search situation suggested. Before optical instruments were available, astronomers used what is called "averted vision" to locate low-magnitude stars, clusters, and nebulae. Henry David Thoreau, who used this method to perform terrestrial as well as celestial searches, raised it to the status of an epistemological principle: "Be not preoccupied with looking. Go not to the object; let it come to you.... What I need is not to look at all—but a true sauntering of the eye" (Thoreau, 1962:488). When you become adept at this, you become "open to great impressions and you see those rare sights with the unconscious side of the eye, which you did not see by a direct gaze before" (592).[3]

Central and peripheral vision generally correlates with the figure–ground distinction. Like all animals that survive by searching for food, while avoiding becoming the food of others, we need to perceive free-standing figures as visually separate from their ambient ground. Figures, once selected, cue the associated images that we animals, bred for the hunt, still carry about in what might be called our internal "field guide"—representations of plants that are good to eat, plants that are insipid or poisonous, animals that are tasty and catchable, animals that are fierce or venomous. Whenever a set of visual data matches up with a set of visual images associated with a known object, that object seems to pop out from the undifferentiated data that surround it (Kosslyn and Sussman, 1994).

This set of distinctive visual features forms a mental prototype that identifies a category of items, such as animals, plants, or otherwise interesting objects. We still use these mental templates in object-recognition tasks, such as picking blackberries, foraging for wild mushrooms, or locating some misplaced item. When language eventually did emerge, these images became tagged as "common nouns," verbal classifications that we store in semantic memory, the cognitive system that makes available to us our general knowledge and beliefs about our world.

Though we can now express this abstract relation of individual to class using this symbolic code, this search process was, and still is, a nonverbal visual skill. Long before language, millions of years before the emergence of our hominid ancestors, animals had evolved the ability to recognize objects by accessing mental images that matched the appearance of real objects in their environment. But the internalized images used in object recognition were not enough. Just as our eye is an organ that not only focuses on figures but also scans the ground in the peripheral field, our brain processes not only imaged objects but also their positions in space—and their movements in time relative to our own positions and movements. When, for example, our prelinguistic ancestors established a camp, however briefly, they needed not only their images of *what* but also the image-mediated knowledge of *where* and *when*: when the gazelles went to drink at the lake, where a grove of fruit trees once glimpsed could be found again, when rival predators were on the prowl, and where the path was that led to the safety of the home camp. For this, they needed an imagination that could draw upon episodic memory to construct a mental map of this territory. As for our near relatives, the primate apes, since they plan strategies, they presumably also form mental images of situations in advance of action (Byrne, 1999).

What sort of map could prelinguistic hominids construct? We can assume with confidence that it would not be the sort of diagram that most of us are used to, a flat territory represented as though seen from the sky—back then, only birds had a "bird's eye view." Our abstract model, in which all objects are simultaneously present, implies that our perception of visible arrays is purely, atemporally spatial, ignoring the fact that our real-world experience of space is one of serially exploring in time a three-dimensional terrain, while remaining inside it. Rocks, trees, caves, and hidden springs being more important to hunter-gatherers than aerial distances, their mental maps of such spaces would therefore have been sequential, space–time models. Even today there are occasions when, lost in an unfamiliar locale, we find it easier to ask for a sequential map: "How do I get to Route 17M?" "You turn left at the next stop sign, go past three traffic lights, make a sharp right at the church, follow the signs to 84 West, get on it and, in about 10 minutes, you should see the exit to 17M."

Without language, our prelinguistic ancestors could not, of course, ask or give these sorts of directions. Their mental maps were purely visuomotor. Having been there before, they followed a set of branching paths along a sequence of landmarks or followed someone who could. Once this series of turns had been learned, they could then navigate them with no more reflective effort than we expend getting around our living quarters, finding our socks, our cereal bowl, our house keys, our way up or down our block—our way, perchance, to Route 17M. Now, if we speakers had been asked by that perplexed driver for the way to Route 17M, we would find ourselves having to fall back on our prelinguistic skills. Forced to convert our prereflective visuomotor routine into a step-by-step, wordless, visualization, we would probably cue this process with some such phrase as "Let's *see* . . . the way to Route 17M . . ." before we could find the words to utter it in the form of a spoken sequence.

The way they—and we, too, for the most part—would navigate an unfamiliar environment would be to parse it into a fixed series of events, e.g., turnings left, turnings right, crossing streams, climbing hills, and so forth. Later, we would be able to sum up the separate events as a whole episode and think of it as "a day's walk out there and back again."

Parsing the Visible *Umwelt*

"Parsing" began as a Latin exercise in which a teacher would point to a word in a sentence and ask a pupil to identify it as a particular *pars orationis*, a part of speech (verb, noun, adjective, etc.), and specify its function as grammatically inflected. This parsing exercise proceeds as a cycle, as shown in figure 4.1. First, a sentence is presented as a unit (A) composed of discrete parts open to analysis. The analytic phase (B) is performed in the serial mode and ends in the complete grammatical segmentation of the sentence (C). The synthetic phase (D) begins as the parsed words are grammatically reassembled into phrases and clauses that are processed in the parallel mode. Finally the complete sentence reemerges as a meaningful unity of parts in complete parallel interrelation (A).

When the word "parsing" is applied to visual perception, it refers to the processes by which the brain takes the colors and shapes that fill the visual field and resolves this array into objects and parts of objects. That this "parsing of the visual scene into a spatial array of discrete objects" ends up producing a "unified percept of the visual world" (Milner and Goodale, 1995:5) is only half the story, however. As David Milner and Melvyn Goodale have argued, visual input systems have evolved, not for their own sake, but to assist animals' behavioral output as they seek food and elude predators. As I proposed in chapter 2, perception and action together form the Master Dyad, essential to all sentient life forms. Since all animals, including ourselves, have perceptual systems appropriate to their particular biological niches, all animals exist in species-specific worlds, or, to use Jakob von Uexküll's (1921) term, different *Umwelte*.[4] An umwelt is more than a particular environment: as he defined the noun, it is the only reality that a given species has the perceptual and cognitive equipment to know. (The German word for an actual physical environment is *Umgebung*.) Thus, for example, the umwelts of bats, elephants, hawks, dogs, whales, bees, and humans, while partly overlapping, are each quite different in the way their perceptual systems sample and represent sound, light, smell, and other physical quanta. This idea reminds us that our specifically human reality is a biological construction formed by our unique evolutionary history and therefore merely one of an infinite set of possible umwelts. In forming *our* world, visual perception has been the determining factor.

I just mentioned that old folk theory of extramission. We may have abandoned that one, but we may have our own unexamined notions of visual perception. One of them, I might suggest, is the assumption that

THE WORLD AS WE SEE IT

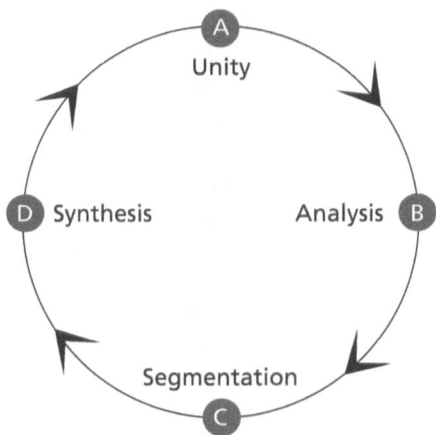

Figure 4.1 The parsing cycle

the total visual field we are aware of when we open our eyes can be accounted for by analyzing the anatomy of the eye. Were visual perception actually completed at this stage, we would not need the brain's intricate tracery of neural paths and way stations. A more accurate explanation for the world as it visibly appears would have to begin by acknowledging that the visible umwelt is the final result of what the entire visual brain does with the photons that enter the lenses of our eyes and fall upon our retinal cells.

The first thing the brain does with the light patterns that excite its retinal cells is transmit these as impulses deep into itself where they are shunted off in milliseconds to be parallel-processed in a number of different areas, each specialized in extracting different kinds of information, such as color, luminance, contours, and motion (figure 4.2). While this bottom-up, feedforward process is going on, other areas of the brain are busy integrating these parsed objects into whole figures and, in a top-down, feedback operation, recognizing them in the context of prior experience. As these processes are occurring, yet other areas are adjusting the irises of the eyes to admit just the right amount of light through the pupils to the retinas, shifting the focus of the eyes in saccades (the speediest muscular reactions of which the body is capable), and repositioning the head to view objects from various angles.

Physiologists have long understood the general direction that visual information takes as it passes from each eye into the deeper recesses of the brain—how it passes along the optic nerves to the back of the brain,

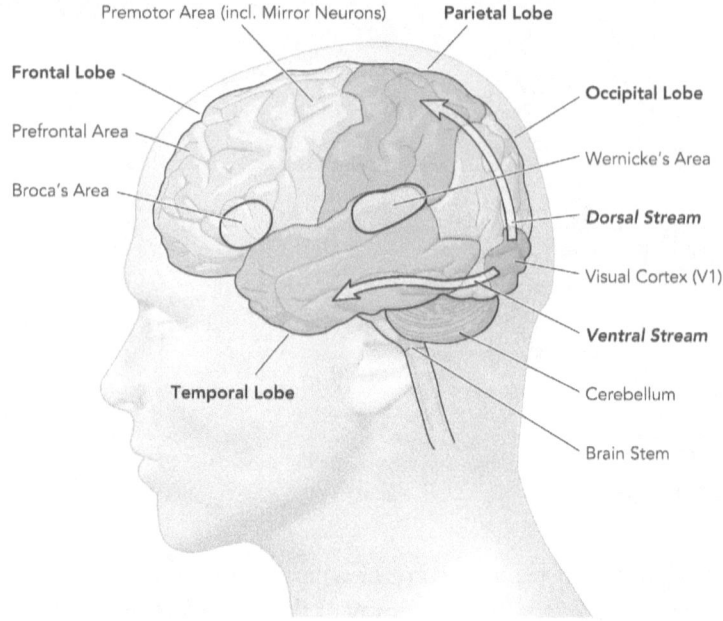

Figure 4.2 The two visual streams in relation to the major areas of the brain

the visual cortex of the right and left occipital lobes, where initial representations of visual data are formed.[5] But not until the 1980s was it understood how the brain subsequently processes this representation. Basing their report on their study of the visual system of the macaque monkey and accounts of patients with particular brain lesions, Leslie Ungerleider and Mortimer Mishkin (1982) identified two neural streams, or pathways, that project forward from the visual cortex. (See figure 4.2, the two arrows.) One stream culminates in the parietal lobe (the upper side region of each hemisphere), the other in the temporal lobe (the lower side region near the ear). The dorsal stream, they said, is associated with spatial perception and might be called the "where?" stream. It is this that uses mental mapping as a guide for moving about and exploring spaces. The lower, or ventral, stream, which they called the "what?" stream, is associated with object recognition (Goodale and Milner, 1992; Jeannerod, 1997, 2006) and is largely responsible for the way our species categorizes the contents of its environment. It is this stream that "transforms visual inputs into perceptual representations that embody the enduring characteristics of objects and their special relations. These representations enable us to parse the scene and to think about objects and events in the visual world" (Milner and Goodale, 2008:774). Recognizing differences that *make a difference*, the ventral stream thus segments the visual array into meaningful components.

According to French neuroscientists Marc Jeannerod and Pierre Jacob, the parsing process of the ventral stream involves "two complementary functions: selection and recognition" (2005:302–3). As they point out, selection sorts out a complex visual array into separate objects, only after which is recognition possible. This is consistent with my biphasic parsing model (figure 4.2): the selection process corresponds to the analytic phase and the recognition process to the synthetic phase. In terming the function of the ventral stream "semantic," they suggest how important preexisting knowledge is to visual perception in that it uses information stored in semantic memory to recognize perceived objects and understand their properties.[6]

For us, the semantic use of visual representations is necessarily modified by language. Unlike our prelinguistic ancestors, we have nouns to help parse our umwelt, nouns at different levels of abstraction. So, we can recognize that gray and rusty blob hopping on the lawn as a "living thing," a "bird," or a "robin," and that greenish formation beyond it as a "tree," an "evergreen," a "conifer," or a "pine." Every *count noun* (a noun that can take a plural, e.g., "flower," unlike a *mass noun*, e.g., "vegetation") can be placed on a scale ranging from the general to the specific—from the basic-level prototype to the exemplar, or token. In the previous examples, the prototypes are "bird" and "tree," and the exemplars are "robin" and "pine." Despite the addition of language, our umwelt is still coded in images, as well as words (Paivio, 1971, 2007).

If the ventral stream manages the categorization of the visual array by first selecting figures and then recognizing them in the context of knowledge stored in semantic memory, we need to consider again the nature of this knowledge. According to Lawrence Barsalou, the data that constitute our knowledge of the world are grounded in the very same perceptual and motor systems through which we first experienced them and are grounded also in the thoughts and feelings that initially and over time have personalized them for us. Everything we know, whether it is a breed of dog, a fruit, a friend's face, a melody, a speaker's accent, a bicycle, a hammer, a swim, a dance step, even a concept such as "travel" or "justice," is stored in its own ad hoc neural network. From each network additional branches connect to areas associated with the processing of kinds of information, including color, contour, orientation, weight, texture, pitch, volume, timbre, smell, flavor, movement, emotion, and so forth. Separate features such as these, abstracted from actual encounters with given entities, stream back into preconscious attention to constitute our recognition of them when we re-encounter them in perception or in thought. Essential to Barsalou's theory is his assertion that our knowledge of the world is not, as narrow cognitivist doctrine would have it, stored in amodal symbols or in some "language of thought." His counterclaim is that the brain's semantic memory preserves a virtually infinite number of "simulators," neural connections that, whenever circumstances require a cognitive response, become reactivated. As revealed by advanced brain-imaging technology, these simulators produce a subconscious or preconscious reenactment, a "simulation" of perception and/or action. When we are aware of this awareness, we tend to explain it by saying that we "associate" x, y, and z with that entity—and it is often some elusive emotional tone, or aura, that we identify. When these abstracted traces merge on a conscious level, we experience them not as separate strands but as a sheaf of features, a single concretized representation, a fully formed mental image (Barsalou, 2009:1281).

In formulating his simulation theory, Barsalou (2008) built on the ecological psychology of James J. Gibson, the work of mental image psychologists such as Allan Paivio, Roger Shepard, and Stephen Kosslyn, as well as on the recent findings of mirror neuron researchers. Not surprisingly, cognitive linguists have become intrigued by the implications of this simulation theory (Richardson and Matlock, 2007). As for the relevance of visual simulation to language and cognitive poetics, I will have more to say about that in later chapters. For now, though, I will return to the question of how initial visual input is processed.

While their peers have generally regarded Ungerleider and Mishkin's findings as groundbreaking, some have questioned their interpretations of the two streams. Prominent among these latter have been David Milner of the University of St. Andrews, Scotland, and Melvin Goodale of the University of Western Ontario. In the early 1980s Milner had begun studying a patient (D. F.) who had suffered a severe brain injury that prevented her from recognizing objects. To his surprise, Milner found that, despite this impairment, she was able to perform visually guided tasks involving objects. Apparently, some part of her visual system still recognized different shapes, although she had no awareness she was making such distinctions. Goodale had earlier researched issues involved in eye–hand coordination in relation to the asymmetries of the brain's hemispheres. Both he and Milner had become fascinated by the implications of the two-stream theory. Very much in the tradition of William James and James J. Gibson, they regarded vision as an evolved means of acting upon the world, not simply forming mental representations of it. Hence their stress on visuomotor functions and the title of their 1995 book, *The Visual Brain in Action*.

In the latter study they outlined the research that had led them to revise Ungerleider and Mishkin's model, concluding that, yes, the ventral stream did access representations that could be consciously used to recognize objects, but that the function of the *dorsal* stream was to guide actions, such as reaching, grasping, pointing, and locomotion, in the context of objects. This stream, which, they said, should not be called the "where?" but the "how?" stream, speedily receives information from the retinas in the form of saccade-driven fixations, a series of momentary "snapshots" that it uses to update the viewer's spatial relation to objects in the visual field. These images, held briefly in working memory, constitute what is called the "optic flow." Unlike the ventral stream, which relies on central vision to select and recognize objects of interest, the dorsal stream has also available to it the full peripheral field. So, what it lacks in acuity, color discrimination, and consciously accessible knowledge stored in semantic memory, it makes up for in sensitivity to peripheral motion and in the speed and skill with which it can guide one's movement within one's immediate environment. And, as the case of D. F. strongly indicated, the dorsal stream guides the subject's actions while concealing its own actions. This is not to say that the dorsal stream is wholly dissociated from other regions of the brain or that the overt actions it oversees are subconscious: it simply means that its *own* visual actions are normally performed below the threshold of consciousness (Wright and Ward, 2008; Braun and Sagi, 1990; Bullier, 2003).

These two streams represent a collaboration of narrow, sharply defined, centralized attention with broad, diffuse, peripheralized awareness. In this division of labor the ventral stream, specializing in fine-grained perceptual representations, selects and recognizes figures, while the dorsal stream, specializing in coordinated action, calculates the relative position of agent and objects in the visual ground. Once again we have an interactive duality that displays the familiar dyadic pattern.

Spatial Frames of Reference

As visual processing subsystems, both streams connect us with our light-suffused umwelt, but, as Milner and Goodale have observed, they do so using different frames of spatial reference. From their study of patients who had been left with only one functioning stream,[7] they found that the ventral stream uses an *allocentric* frame of reference; i.e., it recognizes objects as positioned in spatial relation *to one another*—not to the viewer. As befits a system tasked with object recognition, the allocentric frame represents an object as viewer independent, a token of an enduring type; in accordance to the principle of perceptual constancy, it presupposes that an object's actual size, shape, and color remain unchanged by distance, orientation, or intensity of light. On the other hand, the dorsal stream, as "vision for action," places objects in an *egocentric* frame of reference: from the viewer's central position, the location of objects is calculated in terms of left/right, higher/lower, front/back, near/far, etc., solely in relation to him- or herself.[8] It also adheres to the principle of perceptual constancy, but, while the ventral discounts inconstancies, the dorsal stream treats them as vitally important: apparent size change, for example, can indicate movement toward or away from the viewer, initiated either by the viewer or the object.

While the ventral stream, being "other-centered," views objects as one might through a windowpane, as separable figures arranged on a quasi-two-dimensional ground, the dorsal stream surveys a world of objects and the ground that wholly surrounds the viewer. At the same time, using the third dimension of depth, it calculates the distance between an object and one's whole body, a single part (such as one's outstretched hand), or a grasped instrument (such as a long stick). As we move ahead toward our goal, its peripheral capacity permits the dorsal stream to deal not only with the target object but also to gauge our distance from incidental obstacles that might lie in the way or otherwise interfere with our progress. This computation is achieved through motion parallax, for, as we move, the fixed objects either side of us will seem to move past us at different rates of speed: the nearer will seem faster than the farther. This index of relative distance, by the way, works perfectly well in the monocular vision that characterizes the extreme left and right peripheral fringe, since the rod cells of the retinas are just as sensitive to relative motion as they are to the motion of an object within a stable setting.[9]

These two complementary streams, working together as an optical dyad, construct our visible umwelt by resolving it into those representational dyads, figures and grounds. Every object that interests us out there in allocentric space, once it is selected and recognized as this-or-that by our ventral stream, becomes a closed figure, a gestalt, set forth by the ground that surrounds it. An exaggeratedly selective version of this vision is perhaps what William Blake was referring to when he said: "[M]an has closed himself up, till he sees all things through the narrow chinks of his cavern" (*The Marriage of Heaven and Hell*, plate 14). But, while an object-as-figure is closed on all sides, a ground is *open* on all sides, for, had it closure, ground would be figure. From the egocentric perspective that the dorsal stream maintains, the visual "everything" that surrounds the viewer on all sides is only *seemingly* bounded by the oval visual field with its progressively dim peripheral fringe. If we saw the world only with our dorsal stream, we would find ourselves grounded in boundlessness and, as Blake promised, "everything would appear to [us] as it is—infinite" (ibid.).

The two streams, each with its own spatial frame of reference, hold important implications for any theory of imagination. As a simulation of visual perception, imagination must project its iconic representations upon either an allocentric or an egocentric frame. When we simulate the function of the ventral stream, we imagine an object "out there," in spatial relation to other objects. This observer-neutral allocentric projection changes radically when we adopt the dorsal stream perspective. Now we seem to enter the field of imagined objects as we actively engage with these mind-generated images, rather like the way we encounter figures in our dreams.

The ventral and the dorsal streams operate within yet another, larger complementary pair, the right and left hemispheres of the brain. Though each of the latter presents an anatomical mirror image of the other, each has different specialties. Over their 60 million years of evolution, the primates have been ambidextrous except for the last 2.5 million years, when one branch came to favor the left hemisphere (controlling the right side) when performing finely tuned manual tasks, a tendency that has become a defining feature of *Homo*, the "lopsided ape," as Michael Corballis (1993) dubbed him. While the right hemisphere continued to take in and organize sensory information from the world about it, the left began more and more to focus on details—particular sounds, distinctive visual shapes and colors, ways of recognizing and manipulating objects and of fashioning objects to modify other objects. Undoubtedly the right hemisphere retrofitted its own circuitry over time, but its operations did not compete with the left in narrowly focused perception and finely tuned motor control.

As humans gradually incorporated what Dual-Process Theory calls "System 2" skills, it also retained their older "System 1" skills, and the dyadic pattern that was the result of this collaboration slowly restructured the bilateral brain. The parallel–serial difference became more distinct, the right hemisphere specializing in parallel processing, the left in serial processing. Being adept at integrating parallel input, the right perceives the world holistically and, adept at parallel output, governs motor multitasking. The left hemisphere, specializing in serial input, perceives the world part by part, object by object, and, adapted for serial output, came to oversee step-by-step search, tool use, and, eventually, language.

This division of labor is also evident in visual perception. Figure 4.2 represented the left hemisphere only, but we must not forget that the right hemisphere has its own V1 and its own dorsal and ventral processing streams that function in concert with the left. Each hemisphere has become specialized in a different visual function, the dorsal stream of the *right* hemisphere in visually guiding locomotion in egocentric space, the ventral stream of the *left* hemisphere in the visual selection and recognition of objects framed in allocentric space.

How *Homo* Became Sapient

If we found ourselves transported back in time 1 million years to visit our early *Homo erectus* ancestors, we might anticipate this or that response to this or that interaction and would likely be disappointed, even dismayed, by their choice of behavior. But if we reset the controls on our time machine to 100,000 years ago and visited their East African descendants, we might find them, with or without spoken language, somewhat more amenable—more like us. They would look us in the eye, smile, offer us food and drink, and protect us while we were with them. We would conclude that, somewhere along the way, *Homo* had become sapient.

The traits that made us what we think of as "human," I suggest, coevolved with our visual brain. Millions of years before our recognizably sapient ancestors first appeared, primates had evolved a visual system that became their most dependable link with one another and with the world at large. Over time, other cognitive skills emerged, each based on visuomotor coordination.

An early set of primate adaptations was supported by the mirror neuron system (MNS). This neural hook-up permits a visual perception, mediated by the ventral stream and allocentrically framed, to stimulate motor neurons in the viewer's brain to replicate this action in an egocentrically framed simulation mediated by the dorsal stream—all this without overt action on the part of the viewer (Hesslow, 2002). This wholly internal coordination of the two visual streams, each with its own spatial frame of reference, one allocentric, the other egocentric, must have played an important part in the evolution of *theory of mind*, the belief that others share some of the same mental states as oneself. The two visual streams and the mirror neuron system would also have facilitated *mind reading*, the interpretation of others' covert intentions, and *empathy*, the recognition that others share with us a common set of needs and desires and that social cohesion can sometimes be improved by satisfying them. These visually mediated social skills would have been within the capacity of the bipedal primates that preceded our first human ancestors. "Lucy" (ca. 3.2 mya) and her band of Australopithecines would doubtlessly be capable of theory of mind, mind reading, and empathy. This suite of skills, set in place during the episodic stage, would have been improved upon during the mimetic stage: that community of tool-making *Homo erectus* that we imagined visiting some 1 million years ago would have used them in maintaining their hunter-gatherer culture.

Among the traits that undoubtedly increased in importance during the mimetic stage must have been shared gaze, the use of the eyes to induce others to gaze in the same direction. We suspect it must have been important because our hominid ancestors evolved white scleras surrounding their irises, a directional indicator that is unique among primates. Since the sclera is usually most visible to the right and left of the iris and pupil, its horizontal shifts can be significant for cooperative land-roving hunters and gatherers. For objects on a vertical scale, head movement and pointing could also be put to use. Thus, the eyes that in mind reading were used to pick up covert intentions could now become the means of sending overt messages (Kobayashi and Kohshima, 2001; De Waal, 1982).

The availability of stored knowledge accessed as image schemas is, as we have seen, a requisite for ventral stream processing. The use of an image, stored in semantic memory and allocentrically assembled as a template, speeds the parsing of visual arrays via selection and recognition (Kosslyn and Sussman, 1994:1035–36). When coordinated with egocentrically framed locomotion, mental imagery makes goal-directed *search* possible, and *mental mapping* considerably increases its possibility of success. As I mentioned earlier, the typical mental map is a serial network of landmarks, each a recognizable figure that, once noted, recedes into the visual ground. We now have come to regard landmarks and their targets as allocentrically framed (i.e., selected and recognized) by the ventral stream. But we also understand that, as we direct our gaze and movements toward them, landmarks are also framed egocentrically by the dorsal stream. In our serial navigation through actual three-dimensional space, these landmarks serve us best at the very moment that we turn our gaze-centered, forward-moving bodies away from them and they slide off in the opposite direction into our peripheral field.

Tool use, especially the use of tools manufactured for particular tasks, was also built on the perception/action scaffolding of the primate visual system. As the two serial functions of the ventral stream, selection and recognition normally flow seamlessly from one to the other, but in tool use they are kept quite separate. Recognition may occur when choosing a tool and finding appropriate raw material. It may also occur at intervals during the production of an artifact when one stops to compare the object being shaped to a mental image of it or to a material prototype, as well as at the end when one compares the completed product with such models. This relation of artifact to prototype produces an object-to-object (allocentric) frame of reference. But during the *act* of tool use, the only function of the ventral stream is selection, i.e., the narrowed focus of central vision affixed to a figure—in this case the object being shaped.

In tool use it is the dorsal stream, typically in the dominant left hemisphere, that plays the major role. Here skillful action, enhanced as it is by procedural memory, tends to operate within the worker's peripersonal foreground or the peripheral visual field. As Milner and Goodale (1995) maintain, the dorsal stream is functionally unconscious and does not require focal attention—it is in fact inhibited by such attention. Performed within a strictly egocentric frame of reference, tool use requires automatic three-dimensional calculations of the location of the object not only in horizontal and vertical coordinates but also in depth, i.e., the distance between the tool (say, the hammer) from the object (say, the head of the nail). In short, as a serial motor activity, tool use requires the parallel coordination of both visual streams, the ventral focusing on the raw material and the dorsal guiding the action of the tool user.

The visual system also provided the preadaptive means for the evolution of what are often designated the "higher cognitive" skills. Paramount among these is *imagination*, the voluntary manipulation of mental representations, detached from outward perception and action. For scores of millions of years our mammalian ancestors used images to support and augment perception (Kosslyn and Sussman, 1994). Stored in semantic memory, the features from which whole images are formed could only be activated by online external stimuli. Mental images could be put to other uses, however, and we appear to be the one species able to do so—namely, to access and manipulate such simulations of perception, deploy them in an inner theater of virtual reality, and with them carry on that uniquely human *offline* activity we know as "thought" (Suddendorf, 1999).

The special character of image-mediated thought is its independence from the spatiotemporal present. By means of it, we can visualize persons and settings that lie beyond our current perceptions. Accessing the contents of semantic memory, we can project our thought into the known past, the spatially elsewhere present, and the reasonably predictable future. But without language to mediate these projections, it would seem extremely difficult to sustain image-mediated thought about facts not personally experienced, such as events that happened before we were born or may happen to future generations. Even we, who can entertain such thoughts when considering the past and the future, feel more comfortable projecting our thoughts into our own lived past and our potentially livable future. In short, mental time travel for us is usually experienced as a simulated episode temporally labeled as past or future but having the visuomotor character of an actual present experienced from within our own egocentric frame of reference.

Just as semantic memory could become a resource for the solitary thinker, so too could *episodic memory*. The latter store of time- and place-specific experiences are the result of event parsing, the binding of a series of part-perceptions into a unit that we tag as a separate episode. Since an episode, if it has been stored in long-term memory, must have made an impression on us and aroused us, it is also tagged with a particular emotional tone. When we retrieve it, we place it in an egocentric perspective, one in which we seem to re-experience events in a particular place and time (Tulving, 1983). This involves a sequencing of mental representations, which, like all serial tasks, requires a degree of effort.

Another reason why retrieving episodic memories is effortful may be the fact that the dorsal stream, which mediates three-dimensional egocentric perception for action, is incapable of forming clear, recognizable, enduring images—this is the work of the ventral stream. Though it takes its egocentric frame of reference from the dorsal stream, episodic memory must rely for the detail of its visual representations on semantic memory as mediated by the ventral stream. When we recall an event from our past, its images are recoded in the ventral format and now viewed as discrete figures in a set sequence. What had been initially experienced as an ongoing, wide-angled, parallel-featured episode is now, except for its affective tone, thoroughly serial. The brain, in short, must translate imagery from the egocentric frame of the dorsal into the allocentric frame of the ventral.

As with most translations, something is lost and something else inserted. For this reason, in a court of law, eyewitness testimony and "recovered memories" are routinely adjudged less trustworthy than expert testimony and forensic evidence. Current research into the neural mechanics of episodic memory holds profound implications for the poetics of narrative point of view and imagination, a topic I will return to in later chapters (P. Byrne et al., 2007; Gomez et al., 2009).

Complementarity—The Limits of Human Knowledge?

In the 1990s, the phrase "massively parallel" became a shibboleth among those who were then discovering the similarities between the brain and the computer as information processors.[10] Serial processing, by contrast, connotes a rather unspectacular plodding along of impulses or thoughts. However, serial plodding is sometimes unavoidable, when, for example, one learns a skill, finds one's way in an unfamiliar territory, or encounters new ideas. Since it must be efficient in both modes, the brain might be more accurately described as "massively *complementary*." This is especially true of visual cognition, which is possible only through the complementary (dyadic) relationship of figure to ground, of ventral to dorsal streams, and of allocentric to egocentric frames of reference. Briefly summarized:

1. **Figure and ground discrimination** is essential to the process of selecting, prior to recognizing, the objects we encounter. The visual system uses the serial mode when it performs saccade-driven foveal fixations to determine the boundaries of an object. At the same time, it uses the parallel mode as it preattentively monitors the objects that lie in its peripheral field. These objects belong to the ground by virtue of the fact that their dissociation from the figure defines that object. In other words, the ground generates the figure and the figure generates the ground.

2. The **ventral and dorsal streams** are likewise complementary in their use of serial and parallel modes. The *ventral* cannot broadly scan an array and instantly recognize differences. Instead, it must depend on its sharply focused central vision to fixate on the details that mark each object's boundaries before it can select that object from its ground. Moreover, since it must also depend on semantic memory to ascertain the identity of these objects, it must sometimes double-check itself, a further serial process. The dorsal stream, receiving generous input from the pe-

ripheral field of the retinas, oversees movement through space, a process that involves the parallel cognition of objects, standing or moving, and of the moving body of the viewer. By moving toward objects, it clarifies them, thus serving the purpose of the ventral stream, i.e., the selection and recognition of those objects. Separated as these two visual systems are, they do share information and, different as their functions are, they do, when necessary, actively assist one another. They are, in the words of Milner and Goodale, "different and complementary" (1995:29; see also 1995:53, 177, and 202).

3. The **allocentric and egocentric frames of reference** are the two very different ways we have to locate ourselves and other things in our visible umwelt. The allocentric utilizes the serial mode and does so in alignment with figural vision and the ventral stream. To help us locate where we are relative to our goal, the ventral maps the relation between objects as landmarks. At the same time that it is performing this sort of computation, its complementary stream, the dorsal, will be deriving egocentric information from the same data. In filmic terms, the allocentric combines long shots and close-ups, while the egocentric represents a scene as though taken by a hand-held camera and instantly fast-cut. The egocentric frame needs to be parallel, massively parallel, because, as the viewer moves, it must accommodate objects at different distances in a wide-angled visual periphery. Yet these two frames of reference not only share usable information with one another but do so without ever revealing to the viewer how radically different their separate versions of the world really are.

When we humans began to contemplate ourselves and our world, we did so using neuroanatomy developed over many long ages of vertebrate evolution. This bilateral body, culminating in a bone-encased sensorium, that is in turn wired into a bi-hemispheric circuitry of nerve cells, dictated the conditions under which we came to know reality—our reality, our umwelt. Having inherited this body plan, it seems inevitable that our species would have construed this reality in terms of paired opposites. It still "makes sense" to us to parse an issue "on the one hand" and "on the other hand," and whenever we hear a dichotomy or a polarity offered to clarify a complex or murky state of affairs, it "sounds right," at least at first. I will not claim that bilaterality is the only reason why we tend to think dualistically, but I do assume that one underlying set of opposites, the parallel and serial modes, has furnished us a template that we readily project onto any mass of initially perplexing data that comes our way.

For starters, consider space and time. Space, as it is generally understood, is that entity in which separate objects coexist in parallel at a single moment of time. Time—on the other hand—is that entity or process in which discrete actions or events occur serially in consecutive moments. We may be familiar with the theory that time is the fourth dimension of space, but for most of us this Einsteinian definition of space-time is far easier to say than to conceptualize. Instead, we live our lives as though we still believed with Isaac Newton that space and time are separate and absolute entities. These two concepts, which Kant called a priori intuitions, to us seem so self-evidently real that we entrust to them the governance of our daily lives. But, real or not, these two main coordinates of our human umwelt are, I hypothesize, simply the brain's dyadic structures, its parallelism and seriality writ large.

Sometimes the parallel/serial dichotomy becomes an either/or issue and complementarity seems completely out of the question. Consider the early-twentieth-century controversy between the neuroanatomists who held that the central nervous system communicates by continuous networks of diffuse, filamentary nerve cells (the *reticular theory*, championed by Camillo Golgi) and those who maintained that nerve cells are separate and communicate by electrical charges that leap across gaps, called synapses (the later *neuronal theory*, championed by Santiago Ramón y Cajal). Both sides in that debate agreed on one thing only: the reticular model, essentially parallel, could never be reconciled with the neuronal model with its serial pathways. Consider, too, the more recent controversy between those who see the brain as a set of modules, each dedicated to a single cognitive function, and those who stress the brain's "massively parallel" connectedness. Also, consider how those two terms, "pathway" and "stream," have been applied since the early 1980s to the dorsal and ventral projections: "pathway" implies a conduit for serial locomotion, while "stream" implies a continuous movement of parallel contents.[11]

Then there was Niels Bohr, the scientist who in the late 1920s had introduced the concept of "complementarity." Energy, he maintained, behaves equally as a wave and as a series of particles, a claim that provoked, at the time, a mixture of admiration and dismay from fellow physicists, including Heisenberg and Einstein. On the evening of November 17, 1962, in a taped interview with Thomas Kuhn and several others, Bohr confessed how influential to his complementarity theory had been William James's theory of consciousness. The 77-year-old physicist then recalled how Edgar Rubin, the Danish Gestalt psychologist and his second cousin, had recommended that he read James, especially chapter IX of *The Principles of Psychology*, and he promised his guests that the next day "we shall really go into these things." That promise he did not keep, for the next day he unexpectedly died (Feuer, 1974:126).

What precisely had the father of quantum physics learned from the father of modern psychology? We can only speculate on that, but it is interesting to note that in that chapter James revealed that he, too, had long struggled with two similarly competing paradigms, one parallel and the other serial. For most of the nineteenth century, the dominant philosophy of consciousness was still associationism, as Hobbes, Locke, Hume, and Hartley had formulated it. According to this doctrine, thought was constructed out of "simple ideas," a kind of "mental atoms and molecules" (James, 1890/1950:231). This James rejects as not only empirically groundless but also as inconsistent with his introspected knowledge of thought processes: "Consciousness . . . does not appear to itself chopped up in bits. Such words as 'chain' or 'train' do not describe it fitly as it presents itself in the first instance. It is nothing jointed; it flows. A 'river' or a 'stream' are the metaphors by which it is most naturally described. In talking of it hereafter, let us call it the stream of thought, of consciousness, or of subjective life" (ibid.:240).

Soon afterward, however, James acknowledges that speed of processing changes the apparent behavior of thought. This stream, when it moves at a slow, stable rate, exhibits its "substantive parts," but "when rapid, we are aware of a passage, a relation, a transition from it, or between it and something else. As we take, in fact, a general view of the wonderful stream of our consciousness, what strikes us first is this different pace of its parts." This leads him to switch his metaphor. "Like a bird's life, it seems to be made of an alternation of flights and perchings. . . . The resting-places are usually occupied by sensorial imaginations of some sort, whose peculiarity is that they can be held before the mind for an indefinite time, and contemplated without changing; the places of flight are filled with thoughts of relations, static or dynamic, that for the most part obtain between the matters contemplated in the periods of comparative rest" (ibid.:244). This second metaphor may not be the associationists' sequential "chain" or "train of thought," but, being consecutive, it is no less serial—no less serial than the ocular saccades and fixations that these "flights and perchings" so closely resemble. Together, James's stream and bird-flight metaphors form a complementarity that Bohr must have recognized as a pattern consistent with his own wave/particle complementarity (ibid.:244).[12]

In 1947, when Bohr was honored by the Danish government with admission into the prestigious Order of the Elephant, he was required to

Figure 4.3 Niels Bohr's coat of arms
(From Wikimedia Commons, created by GJo, CC BY-SA 3.0)

place his coat of arms in Frederiksborg Castle. Having no family crest, he had to invent one. He chose for his motto the Latin phrase "contraria sunt complementa" (opposites are complements) and for his image the *Taijitu* (*T'ai-Chi T'u*), representing the two opposite and complementary principles, the yin and the yang (figure 4.3). This pair, enclosed in a circle, have an internally shared common border, a fact that constrains the viewer to assign figural status first to one and then to the other (Arnheim, 1961). As his cousin, Edgar Rubin, had proved with his well-known vase/face experiment, when forms share a common border, each becomes a "reciprocal" image, alternately perceived as figure *and* as ground—a graphic demonstration of complementarity.

The question remains: Is complementarity real or phenomenal? If it is real and not a product of clashing metaphors, then the nature of things is a marriage of radical opposites. Bohr's suggestion that this might be true was what had provoked Einstein to remark that, as far as he was concerned, God did not play dice. Bohr acknowledged the paradox but also understood that, when it comes to the ultimate nature of things, human knowledge must sometimes be content with the phenomenal. The most objective cosmology that our subjective, all-too-human brain is capable of conceptualizing may, after all, be simply another dyadically constructed representation of our human umwelt.

five
Human Communication
FROM PRE-LANGUAGE TO PROTOLANGUAGE

In the preceding chapters I have reviewed some of the basic cognitive skills that were significantly modified during what Merlin Donald has called the episodic stage. During this period, which began 70 mya and lasted until genus *Homo* first appeared (ca. 2.5 mya), the primate brain gradually attained the capacity to convert perceptual and motor events into meaningful units of experience. Attention, no longer restricted to a moment-to-moment window on the world, could widen and take in more and more details, compare them with stored memories, and do so over longer intervals of time. Rather than being bound by instinct to respond to parallel-perceived stimuli with parallel-coordinated responses, primates could at times use their widened window of attention to observe serially enfolding events and respond to them with serially ordered actions. The window metaphor is especially appropriate, since it was their acute visual perceptual system that provided our primate ancestors, both in their arboreal and their terrestrial habitat, with their most reliable environmental information.

Being social animals, primates had additional skills, such as theory of mind, mind reading, and empathy, plus a variety of gestures and cries used to convey their wishes and emotions. It was upon these social skills that more discriminative forms of communication were later built, adaptations that ushered in the mimetic stage and with it genus *Homo* and the Paleolithic era. Donald's (2007b) apt neologism, "mindsharing," applies *mutatis mutandis* to all four of his stages: to prehuman primates, as well as to archaic and modern humans, to unintentional signs, such as gaze and emotive facial expressions, and to intentional signs, such as gesture, speech, and writing.

That the evolution of mindsharing should have culminated in language was not, however, a foregone conclusion 6 mya when our branch separated from the main primate stem. It did happen, but when that enhanced mindsharing first began and what it was like remain matters of ongoing controversy. Before I sketch out some of the arguments in the contemporary debate, I must report that in the battle between those who favor a sudden leap from nonlanguage to true language and those who favor a gradualist account, the latter seem to have turned the tide. As the preceding chapters indicate, my own view of language is that it is a medium not merely capable of expressing nonverbal cognitive processes, but it is itself the expression of these processes. Accordingly, verbal artifacts, or poems (broadly defined), are able to provide glimpses into the nonverbal embodied mind because they are themselves the consummate instruments of that mind.

In this fifth chapter, I will examine the forms of communication that, most evolutionary anthropologists assume, preceded the form that all humans throughout the world now use—full language with its extensive lexicon and rule-governed syntax. After outlining some counterfactual scenarios in order to demonstrate how un-inevitable language evolution was, I survey some of the recent theories that try to describe the minimal level of communication that a highly social, tool-making, hunting/gathering genus of primates would require. While continuing to use expressive vocalization, this *pre-language* would have expanded the store of manual gestures far beyond that of modern apes.

I then consider *protolanguage*, the symbolic code of syntax-less speech composed of clearly articulated phonemes that many assume had to have been a transitional phase between pre-language and full language. When protolanguage and, later, full language emerged, these retained features of the prelinguistic system, deploying that older repertoire of voice and gesture as *paralanguage* to convey a broad range of affective states and semantic nuances. Those primitive elements continue to accompany speech and, as I will suggest in my next chapter, have become embedded in the medium of literature and to a special degree in poetry, providing it with its traditionally recognized structures.

Why Language?

The evolutionary process builds on the structures already in place. When innovation is called for, it is these structures that must be adapted. But what those innovations turn out to be can never be predicted by simply examining those preexistent structures. The proverbial Martian observer, had he been present, would have found the hominid turn toward orally articulated language as one such unforeseeable innovation. So, lest we fall back into that old assumption that all the changes that led the first apes to become erect, hairless, fire-taming, flint knappers inevitably culminated in language, I will offer three brief thought experiments. Let us imagine that, following the evolutionary spurt that led to genus *Homo*, some 2.5 mya, this new genus did not expand its territory or launch those migrations that brought it into Asia and Europe. Instead, this genus, about 2 mya, proceeded to carve out several conservative niches for itself on the African continent. Then imagine these three alternative niches occupied by three separately evolved species:

    1. *Homo avis*, or "Bird Man." After half a million years of scavenging on the open savannas, fighting off hyenas and running from big cats, some early humans returned to the trees. There, they wove nests and walkways high in the forest canopy. Their arms again grew longer than their legs. They lived in families of from 10 to about 15 persons. Their tools were wooden. They ate nuts, fruits, and birds' eggs, the latter gathered from a species of birds that they had successfully domesticated. Their physical contact with other families was rare and for mating purposes only. When they communicated, they used a sort of vocalise, trilled melodic phrases that signified a very brief number of events, such as the sighting of snakes or eagles, or changes in the weather.

    2. *Homo fodens*, or "Mole Man." These hominids devised tools to dig underground burrows, where they lived in large communities of up to 200 persons. For food they ate termites, rodents, and roots. They eventually learned to farm large insect grubs. Without light, they were functionally blind, but their ears were relatively large and able to be directionally twitched to pick up the narrow range of sounds transmitted in their subterranean habitat. When they approached one another in their corridors, they recognized kin by smell and could communicate intentions by touch.

    3. *Homo aquaticus*, or "Merman." This species moved from the grasslands to the shores of rivers and lakes and became powerful swimmers

whose speed was enhanced by loss of body hair, while layers of subcutaneous fat helped them retain heat in chill waters. They fed largely on fish, mollusks, and wetland tubers. They congregated in herds and came to land only to sun themselves and mate. Their communication was through displays of dominance and submission.

Of the three species, only the last (minus the walrus traits) resembles a scenario that has been seriously proposed. First enunciated by the marine biologist Alister Hardy (1960) and promoted thereafter by the writer Elaine Morgan, this was proposed as a phase through which our African forebears passed on their way to becoming *Homo sapiens*. Though the "aquatic ape theory," as it is called, has not been widely espoused, a number of evolutionary thinkers, including Frans de Waal, Daniel Dennett, and Richard Dawkins, have declined to rule out the slim possibility that hominid evolution may have included this phase.

Had our remote ancestors taken the evolutionary branch in the road that led to any of these three "what-if" scenarios, it is unlikely that they would have developed the communicative skills associated with language. They would not have *needed* to do so. "Bird Man" would have lived in a restrictive space and in such small units that a mere handful of gestures and a minimum of sounds would have sufficed to exchange intentions. Blind "Mole Man's" habitat would also have lacked the variety of input that would require a large vocabulary. In such straightened circumstances there would be little opportunity—or need—to plan or coordinate activity (as for communal grub farming, that could be automatized, as is the aphid farming of ants). Finally, "Merman," who would hunt in small groups swimming below the surface, would be unable to utter vocal sounds underwater or engage in gestural discourse much beyond facial signals and finger pointing.

The consensus scenario, of course, is still the one in which hominid apes, probably the Australopithecines (ca. 4–2 mya), descended from the trees and ventured further and further out into the open grasslands that appeared at the close of the Pliocene epoch. Their descendants, from *Homo habilis* to *Homo erectus*, became omnivores with an increasing appetite for meat and the quick energy it afforded. For this they came to depend on the skill of hunters who had learned to work together, share their kill, and bring back portions to share with the women and children. This is classified as the Lower Paleolithic Age (ca. 2.5 mya to 100,000 B.P.), which Merlin Donald has called the mimetic stage, a time when humans developed and transmitted to their young such technical skills as botanical lore, tool making, hunting methods, and other forms of cultural knowledge. Since the survival of the group depended on a sharing of labor and its material benefits, natural selection favored good communicators (Mameli, 2001).

Nature, already the mother of necessity, thereupon became the grandmother of invention. But why *this* invention? The two most plausible answers to that question are (1) to share *object information* and (2) to share *social information*. Of course, now we use language to do both and assume that both were important factors in human cultural evolution, but which of the two needs most drove the development of the first forms of human communication?

If we consider object information as the primary purpose, we can link signs, be they gestural or vocal, to those sets of visual features we carry about in our brains to identify a perceived object as the token of a type. This capacity, which we share with mammals generally, is the basis of that rich set of skills we know as "imagination." Our prehuman ancestors had undoubtedly used this mental imagery in parsing their immediate environments, but early humans found a new use for them: they could move these whole images or sets of features about and connect them in various ways to generate offline cognition, i.e., thought. Eventually they would find ways to translate this mental pictography into separate gestures and/or sounds and for the first time converse with others about objects *in the absence of those objects*. Once they had mastered this powerful tool, humans could come together to plan for the future.

There was a problem, though, a *social* problem. The first scrap of object information ever communicated was likely to have been a lie. According to "Machiavellian primate politics," only seeing is believing, so information concerning absent events should not be trusted (De Waal, 1982). By contrast, involuntary emotional indices, relatively hard to fake, are much more believable than facilely transmitted signs. As Emily Dickinson wrote, "Men do not sham convulsion, / Nor simulate a throe." But if one's audience always demanded instant verification and one's credit always required such immediate emotional collateral, an informant could never, as Chris Knight (1998) put it, "refer to phenomena beyond the current context of here-and-now perceptible reality" and could never "express a fantasy, elaborate a narrative or specify with precision a complex thought." We have obviously learned to do so, but if no one could "suspend disbelief even momentarily" (Knight, 1998:82), language and its artifacts, including the poetry of Emily Dickinson, would have proved an absolute impossibility.

Despite what lingering primate reservations they may have had, our ancestors did learn to communicate through gestures and sounds because in their expanded environments they needed to communicate their whereabouts, intentions, and perceptions, so those best able to share truthful information were most likely to survive and pass on to others this useful trait. Prelinguistic mindsharing helped everyone. Sharing the knowledge of how to harden a spear point with fire or the news about what lay ahead on the path would be, like sharing food or physical labor, an investment that would be reciprocated by others within the group. Likewise, if one failed to reciprocate a gift of food or assistance or refused to share information, this action would have been, like stealing or lying, an item of information that also needed to be shared with others. Though an offense may have been committed against only one individual, any antisocial act threatened the solidarity of the group and its revelation was yet another motive for information exchange.

The evolutionary path to language required a level of solidarity that Chris Knight (1998) has called "intense ingroup trust." Only when individuals came to recognize that their own survival depended on their group's survival could mindsharing codes be created and maintained. Since "ingroups" presuppose the existence of "outgroups," any sign-system maintained within one group would operate as a secret code, not readily shared with other groups. This followed from the probability that early hunter-gatherer societies, like some still surviving, lived relatively isolated from one another within geographically small linguistic communities.

The fossil record tells us that, over the millennia since the human genus first appeared (ca. 2.5 mya), human communities grew in population numbers and, as they did, so also did the human brain grow in size. The prehuman *Australopithecus* had a brain about the size of a chimpanzee, *Homo erectus* had one almost the size of ours, and, according to some estimates, the median size of Neanderthal brains exceeded our median by as much as 100 cubic centimeters. According to Robin Dunbar, this increase was a response to individuals' need to manage social relationships. Citing current primate ape population density data (e.g., chimps live in groups of about fifty), the fossil evidence, and current human populations, he and his colleagues have estimated that *Homo erectus* groups comprised an average of 111 persons who knew and regularly interacted with one another. The number of archaic *Homo sapiens* was 131, Neanderthals, 144; for modern humans, *Homo sapiens sapiens*, it was and still is approximately 150 (Aiello and Dunbar, 1993). With increased brain size came increased capacity—and need—for the transmission and social archiving of coded thought in some linguistic, or language-like, format.

Like chimpanzees and preverbal infants, early hominids would have first used signs to get others to do what they wanted—to give them nourishment, warmth, comfort, protection, etc.—in the here and now. These manipulative wishes would be mediated by facial and manual gestures and accented by emotive vocalizations. As hominid societies gradually transitioned from hierarchical structures dominated by alpha males to egalitarian systems of hunter-gatherers, they found that information exchange, a new kind of mindsharing, could strengthen ingroup trust and convert sign sending from a potential instrument of deception to an instrument designed to unmask deceivers and castigate cheaters, hoarders, free riders, and other abusers of communal trust. As a pressing motive for communication, gossip became a practice that was to drive the subsequent evolution of both moral and linguistic codes. Unlike simple behavior manipulation, gossip would require participants to form representations of past events, empathize with the aggrieved, condemn the wrongdoer, and admire the skillful, strong, and generous. This form of social discourse, the ur-form of narrative, might have used indexical gestures to draw attention to third parties when visibly present, but it would also need symbolic signs, prelinguistic proper nouns of some sort, gestural or vocal, to identify these persons in their absence. Gossip, as Robin Dunbar (1996) reminds us, continues to be a means of regulating the behavior of antisocial individuals and of preserving social cohesion among the disapprovers.[1]

The need to share both object information and social information must have strongly motivated the development of early forms of communication. Though distinct from one another, each kind of information is rooted in the perceptual here and now. Object information derives from object recognition, that instant when a mental image matches up with an *actual* visual percept, and social information is enacted in the form of a plea, an offer, a command, a threat, etc., within an *actual* situational context. Though anchored in the perceptual present, both kinds of information also rely on long-term memory. Object information relies on semantic memory to verify sensory input, while social information relies on episodic memory, albeit generalized, to determine to whom one may safely direct a plea, offer, command, etc.

As online knowledge, these two kinds of information provide an individual with the possibility of thought, examples of which, if verbalized, might be "Ah, this is an egg (not a stone)" or "I can be protected by *her* (because she doesn't like *him*)." It was as *offline* knowledge, however, that these two kinds of information had to have most challenged the cognitive powers of prelinguistic humans. Thinking and then sharing such knowledge with others in order to survive in an uncertain environment required them to take their knowledge of the present, in which all things exist in parallel, and transpose it to some absent world in which only serial events are thinkable—a temporal *once*, a spatial *elsewhere*, and a temporal *not-yet*. Both social and object information could now be shared as a communal archive of cautionary tales, useful in curbing deviant behavior and promoting virtue and also as a broader narrative of ancestors, culture heroes, gods, and personified powers of nature that in the present inhabit the spatial elsewhere. Projected into the temporal not-yet, mindsharing took the form of planning, the essential human activity that the Greeks personified as Prometheus, god of foresight and maker of mankind.

Evolutionary scientists, when they speculate on early hominid culture and its role in the emergence of language, often speak of object information in terms of foraging and ecological know-how and of social information in terms of interpersonal trust and the intelligence required to manage complex divisions of labor. Derek Bickerton (2002) expressed his view that the advocates of social intelligence had overstepped the bounds of evidence when they claimed that language emerged as an effect of social intelligence: no, it was language that had been itself the *cause* of social intelligence. Any pre-language would have emerged out of an absolute need to know the *non*human environment, e.g., available food sources and the dangers posed by predators. Information exchanges, according to Bickerton, "need not have been monomodal, nor need their units have been arbitrary in the Saussurean sense. Directional gestures with the hand, accompanied by the imitation of the noise of a mammoth, could easily have been interpreted as meaning 'Come this way, there's a dead mammoth'" (2002:219).

But Bickerton's example of ecological intelligence fails to disprove the social intelligence hypothesis. While it is true that simply knowing how to behave in a socially appropriate way may not have been as urgently needed as locating a scavenged slab of mammoth, those two needs, rather than opposed, are here quite complementary. Using an indexical sign (pointing) and an iconic sign (the imitated bellow) to communicate that mammoth meat is available would have been a behavior that preserved hominid social solidarity through a sharing of resources, which after all would have been the only purpose of this communication. If sharing social information had not been an overriding moral obligation, the object information that Bickerton here describes would simply not have been transmitted.[2] If we apply a dyadic analysis to Bickerton's example, we can place object information in the central figural position, since for both groups of scavengers the physical fact of the dead beast was the focus of the message. But in the peripheral position, the transmission was grounded—surrounded, as it were—by the general understanding that the community should share the benefit from this find.

From Pre-Language to Protolanguage and Beyond

Having outlined the two most likely motives for communication and opted for the social intelligence hypothesis as the more comprehensive of the two, I will piece together a narrative beginning at the close of the Pliocene epoch (5.3–2.6 mya). Though cooler and less stable than the preceding Miocene epoch, the Pliocene climate had been warmer and moister than it is now, but by its close an irregular trend had already set in. In Africa, forests began to shrink, and grasslands and patches of desert appeared. *Australopithecus*, the bipedal genus of hominid ape that had evolved in the middle of the Pliocene, was still holding its own, competing with other carnivores for food along the wooded borders of the savannas. Unlike the older-evolved quadrupedal and tripedal apes, whose faces were lowered when walking, Australopithecines would have been able to make eye contact and meaningful facial and manual gestures while actively navigating this new environment (MacWhinney, 2004).

A major split in the hominid lineage occurred at the close of the Pliocene and at the onset of a new geological period, the Pleistocene epoch. Some Australopithecine stock apparently returned to the forest and, through natural selection, adapted their jaws for crushing nuts, roots, and the occasional scavenged bone. These became the *Paranthropus* genus. The other Australopithecines stayed in the competition for herd animals, gradually reduced the size of their teeth and the power of their jaws, and learned to fashion stones into hand axes to serve as external teeth with which to split and crush the food they found. These eventually formed a new genus, *Homo*, the earliest identified species of which has been called *Homo habilis* (2.5–1.4 mya), succeeded by the related species *Homo ergaster* (1.9–1.4 mya) and *Homo erectus* (1.8–0.2 mya). Being tool makers, these species have been designated the first human hominids and, to the extent that serial (sequential) aptitude is preadaptive for information exchange, may have been the first to string together signs, gestural and/or vocal, to communicate their thoughts.

As I have noted, Bickerton's example of the communication between scavengers on the open grasslands uses a visual index and an auditory icon. As an indexical sign, pointing is a typically human gesture. Though some chimps in captivity have learned to do so, no other animal can use or interpret the pointing gesture as a way to direct attention elsewhere. Like that visual index, the auditory icon directs the perceiver's attention not to the sign sender or to some actually bellowing mammoth but to a mammoth that, in this context, is no longer dangerously alive. Though language would utilize symbolic signs in the form of arbitrary, conventional vocal sounds, the pre-language of early hunter-gatherers must have featured a bimodal repertoire of indexical and iconic signs adequate to the needs of Paleolithic society for over a million years before fully symbolic signs were devised.

Up until the 1990s it was assumed that *Homo erectus* and related species, lacking true language, could not plan big game hunts but had to depend instead on scavenging carcasses left by larger predators. They were, it was thought, clever bipedal omnivores that, armed only with clubs and sharpened hand axes, roamed in packs, like their main competitors, the hyenas. According to this scenario, it was not until anatomically modern humans, *Homo sapiens sapiens*, evolved (200,000–160,000 B.P.) that hunts could be adequately planned and executed, first with the aid of an advanced protolanguage, later with a fully grammatical language.

This view, however, had to be reevaluated when in 1995 Hartmut Thieme uncovered and identified seven wooden spears in Schöningen, Germany. Their examination of this find led researchers to infer that these finely crafted, balanced, aerodynamic throwing spears, stratigraphically dated ca. 400,000 years ago, were manufactured for hunting parties to use on large game, such as horses, the bones of which were found among these artifacts. If some form of symbolic communication was necessary to coordinate such hunts, *Homo erectus* or some related species, such as *Homo heidelbergensis*, had to have used some sort of pre- or protolinguistic code a good 200,000 years before our own subspecies evolved in Africa. This archaeological find further suggests that Acheulian blades, the bifacial design of which seems not to have changed over 1.5 million years, were principally used to sharpen other, less durable materials (e.g., wood and bone) and that the latter constituted the primary raw material for these craftsmen (Mithen, 1996:96). Lest we think of pre-*Homo sapiens sapiens* humans as lurching about, grunting, and otherwise killing time until we nimble, fast-talking cave painters first issued forth from Africa some 90,000 to 70,000 years ago, we should not forget that it was *Homo erectus* that undertook the first great human migration that by 1.5 mya had brought our genus from Africa through Asia and into Europe, venturing as far north as the British Isles (Wade, 2010). To have survived for 1.5 million years in a multitude of climates and terrains, they must have had the capacity to work together to achieve social objectives. For this, communication skills would have been required.

When anatomically modern humans appeared in East Africa 200,000 years ago and began to spread, they were but one of several hominid species then inhabiting that continent. When by 90,000 years ago a growing population of *Homo sapiens sapiens* began expanding into the Middle East, into southern and east Asia, northward through the Caucasus, and by 45,000 years ago into southern Siberia and Europe, they encountered older human species, offshoots, like themselves, of *Homo erectus*. Notable among these was *Homo neanderthalensis*, a species that had first become adapted to life in Ice Age western Asia and Europe. The newcomers no doubt had other skills that proved them superior to this and other older species, but most advantageous among these must have been fluent, syntactically sophisticated language.

Paleoanthropologists have come to refer to our subspecies as *Homo sapiens sapiens* to differentiate it from the less advanced "archaic" *Homo sapiens*, which, as some maintain, may have included *Homo neanderthalensis* and *Homo heidelbergensis*. This new subspecies to which all living humans now trace their genetic origins had a vocal apparatus able to produce a wide range of distinct consonants and vowels as well as the motor control able to do so with prodigious speed. Moreover, its perceptual and information processing capacities were equal to the task of distinguishing the meaningful elements in this cascade of sound. Even if an older species, such as the Neanderthals, had vocal language, they could never have communicated among themselves at the speed, not to mention the semantic and syntactical precision, of *Homo sapiens sapiens*.[3]

But if so, what sort of pre-language and protolanguage could pre-sapient humans have possessed? Gesture, which we still fall back upon while trying to communicate across languages, would certainly have been used, augmented by imitative sounds, as Bickerton illustrated. These gestures and sounds would, over time and within a given community, become "conventionalized," i.e., assume abbreviated abstract forms (MacWhinney, 2004; Burling, 2000; Corballis, 2002). If so, the process of change would resemble that demonstrated by the Greek alphabet: the

Canaanite/Phoenician symbols, first borrowed from iconic hieroglyphs for ox, house, hook, etc., only later became fully abstracted as the Greek signs for alpha, beta, gamma, etc. Eventually, such gestures and sounds could have become so abstracted as to become conventional symbolic signs, as are almost all the words in the lexicon of every modern language.[4] Still, the means needed to produce this gestural pre-language would have been less efficient than fluent speech. Gesturing would interrupt manual labor and could not be used in darkness, while imitative vocalizations, as such, would have limited referentiality unless they were radically abstracted, in which case they, too, would become the symbols of a conventional lexicon.

This latter possibility is associated with Bickerton's early work, which culminated in *Language & Species* (1992). Strongly influenced by Chomsky's "catastrophism," Bickerton declared that at some time between 290,000 B.P. and 140,000 B.P. there occurred, "an event, presumably a mutation of some kind, that affected a single female living in Africa" (1992:165). Up until then, humans had communicated through gestures and sounds and eventually through an orally produced protolanguage of basic nouns and verbs, a pidgin, such as that still used today among communities that do not share the same grammatical language. As a consequence of this brain-changing, speciating mutation, "syntax must have emerged in one piece, at one time," and this woman ("Mitochondrial Eve") and her descendants became the first members of our subspecies *Homo sapiens sapiens* (1992:190). Pidgin, as a language form typically used to negotiate trade among strangers, stresses object information, a fact that may have influenced Bickerton's belief that the precursors of *Homo sapiens* were not concerned—or unable—to communicate self-reflexive social information.

Another possible form of protolanguage, one that better fits a social information purpose, is the formulaic hypothesis that Alison Wray (1998) has proposed according to which early speakers (from *H. erectus* to archaic *H. sapiens*) communicated among themselves in single, semantically unsegmented phrases, which she has called "holistic," or "formulaic," utterances.[5] Rather than suppose a system of separate sounds representing separate agents, actions, and objects, Wray envisages a simpler form aimed at conveying social information alone—i.e., getting others to do, or *not* do, specified actions. This protolanguage may have been the behavior that, as Dunbar (1996) argued, would have taken the place of primate social grooming as a means of maintaining peaceful relationships among group members. At any rate, we modern humans have inherited this universally practiced discourse genre from some ancient source. Consider, for example, the function of some of our own formulaic phrases: "Beg pardon," "How do you do?" "My pleasure," "You're welcome," and "Take care." These idiomatic phrases cannot be broken down into component parts or grammatically altered. Knowing what *pardon*, *pleasure*, *welcome*, and *care* mean as separate words does not help us understand the meaning of these social formulae, nor can we preserve the function of "How do you do?" if we transpose this greeting to another tense or person (e.g., "How did you do?" "How do I do?" or "How will he have done?").

This protolanguage would have been accompanied by gestures and situational contexts, but there would be no grammatical context: "Each utterance," according to Wray, "would have been stand-alone, and devoid of any internal structure." To illustrate her theory, she imagines the utterance "mabu" as used to mean *keep away*; "madu" to mean *take the stick*; "mebita" to mean *give her the food*; and "ikatube" to mean *give me the food*. Beyond such social-interactive communications there would be nothing sayable, perhaps nothing even thinkable. "There would be no reference or description for its own sake (*this is a tree; the tree is tall*)" (1998:51). There would be no way to say "*the world looks lovely tonight* or *I wonder if it will rain tomorrow* or *it wasn't like that in my day*" (Wray, 1998:51–52, author's emphasis).

For language to have grammatical, rule-governed context and utterances to have internal structure, vocal sounds would have to be segmented into individual words, and rules would have to be devised to arrange these units in patterns that meaningfully reflect the perceived actions and relations of objects in the world. Why, after perhaps a million years of speaking holistic protolanguages, should some communities of genus *Homo* have begun to insert single, functionally different vocal signs into their speech? Unless at this point one opts for genetic mutation, one is likely to continue the search for an adaptationist answer to this adaptationist question: What sort of selective pressures might have led some humans to modify their linguistic medium?

In 2007, Wray, in collaboration with George W. Grace, proposed an elegantly simple answer to that question. Language has two different uses: one is to ease everyday social interactions among family members and close associates (esoteric language); the other is to exchange object information and negotiate with outsiders (exoteric language). Esoteric language, with its formulaic commands, requests, and greetings, was, and still is, the "natural default setting for human language" (2007:543).

In communities that, a million years ago, rarely exceeded 120 members (Dunbar's estimate) and rarely, if ever, engaged with other communities, esoteric language would have been the only form ever spoken. Each human community would constitute what Talmy Givón has termed a "society of intimates, where all *generic* information is shared" and, therefore, communication would be "about the *immediate context*, where all specific information is shared" (1979:297). It also follows that early communicative systems could be used as an argot for the purposes of "conspiratorial whispering" in the presence of outsiders (Knight, 1998). Until some other linguistic means could be devised, whenever members of different inward-facing communities did meet, their only available lingua franca would still have to be indexical and iconic gestures.

On the other hand, "exoteric communication is outward-facing and conducted with strangers" (Wray and Grace, 2007:551), and, to avoid misunderstandings, it must be as transparent as possible. To accomplish this, speakers would have to treat utterances as though these could be meaningful when detached from ingroup knowledge and from any specific situational context that may have prompted them. Such utterances would have to use sounds—i.e., *words*—that could be autonomous and interpretable in isolation (555). Such requirements would lead to the invention of rules that allow speakers to refer to temporally displaced, conditional, counterfactual, and as-if mental spaces, domains that no holistic protolanguage could ever access.

Wray's theory presents some other interesting implications for the emergence and extraordinary success of our own species. Communities of *H. sapiens sapiens* and possibly other related species grew in population and became acquainted with one another (traded, intermarried, engaged in cooperative ventures). For this they needed to use a common lexicon and syntax alongside their own esoteric, formulaic vernacular (ca. 200,000–100,000 B.P.). The more enterprising of these communities took this common exoteric language with them as they ventured beyond their East African homeland. As they expanded their range and encountered scattered human communities, those indigenous people were obliged to learn this new code in order to interact successfully with these newcomers (ca. 100,000–60,000 B.P.).

Those communities of *H. erectus*, *H. heidelbergensis*, *H. neanderthalensis*, and *H. antecessor* that could parley with the newcomers could form alliances with them in order to drive off or kill their local rivals and take their plant-foraging and hunting grounds. Thus, this new means of communicating not only social information but also, and most importantly,

HUMAN COMMUNICATION

object information could spread outward as items in a complex, self-replicating meme. Those that could not learn this or any exoteric language were relegated to their inward-facing holistic protolanguage and over time proved unable to adopt novel ways to confront novel situations. Whether disease, genocide, inability to compete for food, or genetic

TABLE 5.1 A timeline of the major human species and their evolving modes of communication

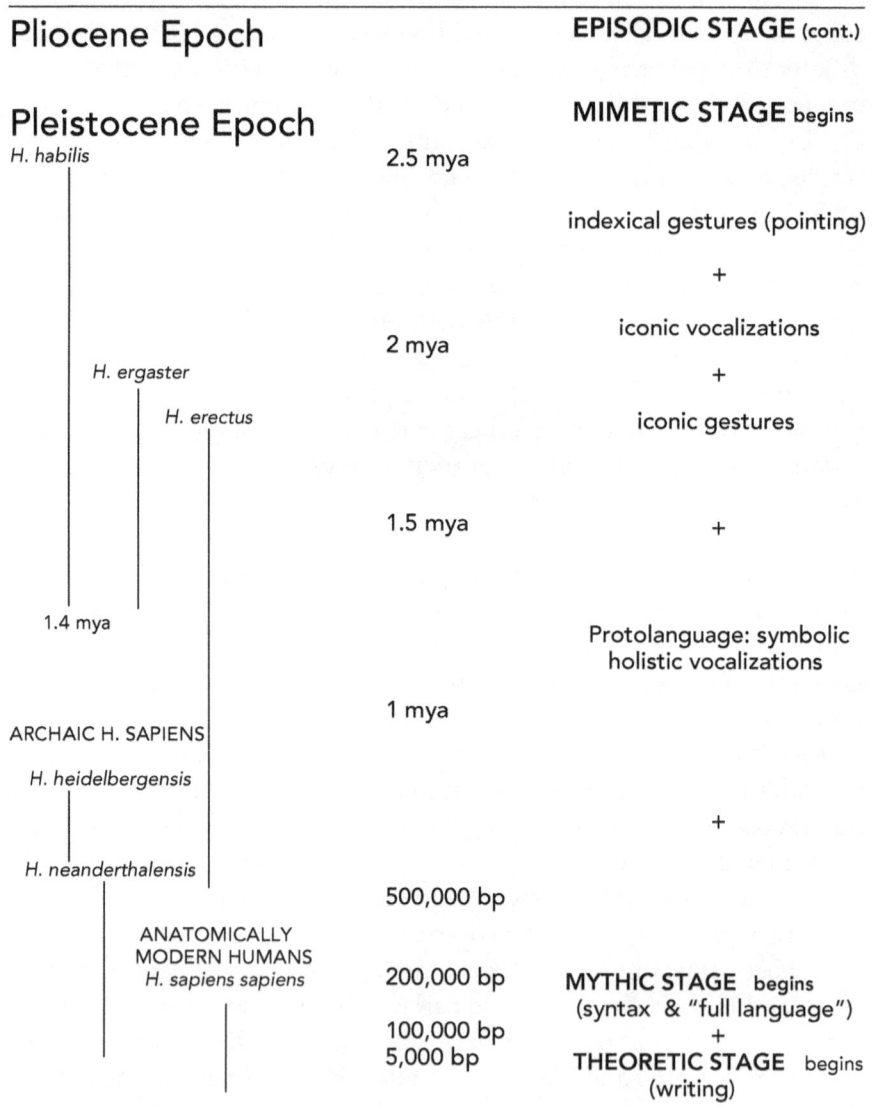

assimilation were factors in their disappearance, their linguistic disadvantage could have been crucial.

The timeline in table 5.1 is meant to embody the major phases as hypothesized by the sources I have cited. The vertical lines to the left represent the duration of species from earliest to last appearance, but they are not intended to show genetic derivations. The plus signs in the right column indicate that these communicative features are additive, not substitutive.

I will now proceed to examine more closely the properties and communicative uses first of gestures, then of voice, and suggest that the transition from pre-language to protolanguage could have been exceedingly slow.

Gesture: Index and Icon

Nonhuman primates have little voluntary control over their breathing and vocal muscles (Hewes, 1973), so the last common ancestor (LCA) of humans and chimpanzees, some 6 to 7 mya, must also have lacked the neural wiring necessary to produce more than a few stereotyped vocal signals. Yet all primates are manually dexterous and manipulate objects using both power and precision grips. Moreover, chimpanzees, both captive (nonenculturated) and wild, have been observed using from 20 to 30 distinct manual gestures together with a number of orofacial gestures (Call and Tomasello, 2006). This combination of vocal deficiency and manual dexterity strongly favors gesture as the earliest human medium of communication. The gestural communication skill of *H. habilis*, *H. ergaster*, and *H. erectus* must have been vastly greater than *Australopithecus*, since their tool-enhanced hunting-gathering life on the savannas would have required planning and coordinated execution. Promethean fire skills they had, but, more importantly, they had to have had Promethean forethought as well.

There are other reasons to infer that our human ancestors relied on visual gesture before speech. First, the area of the left brain hemisphere in modern human and nonhuman primates that is specialized for fine motor actions overlaps with Broca's area, which in humans is specialized for vocal articulation. This fact has led an increasing number of neuroscientists to speculate that speech subsequently evolved when the circuitry used in fine-tuned manual tasks was replicated in the adjacent motor areas that control the muscles of the larynx, tongue, and lips (Iacoboni,

2005; Corballis, 2011). Second, insofar as ontogeny is relevant to phylogeny, the developing brain of an infant is capable of gestural communication months before it is able to communicate its needs through words. And third, needless to mention, when we speak, we persist in gesturing with arms, hands, lips, eyes, eyebrows, and posture, a universal behavioral set that suggests how deeply embedded these mechanisms are in our communicative toolbox.

In the last chapter I discussed some of the ways by which the ventral and the dorsal visual streams determine our species-specific umwelt. While the actual physical environment exerts an overriding influence on the evolution of any animal's neuronal structure, each species, within its own biological niche, is responsible, so to speak, for customizing this equipment. Our own human niche has been characterized by social cohesion, imitation, cooperation, and division of labor, traits that have led us to pool our knowledge of the physical environment, its affordances and dangers. This biocultural niche, an updated version of our primate umwelt, is that spatially and temporally distributed web of objects and events, figures and grounds, parts and wholes, and causes and effects that together compose the "world as we see it." We feel at home in this world because it seems to conform to our internal representation of it and, since this is the only world we know, it is the only world we can knowingly share with one another. Not surprisingly, the various means we have to externally project this knowledge replicate the brain's visual means of representing this knowledge internally. Gestural or vocal, the signs we have for things represent them at the same level of abstraction as mental images, i.e., as context-free prototypes.

Before we can consider how we share knowledge, we need to consider how we *get* knowledge, and for that we must review some of the elementary principles of semiotics. To qualify as a sign an object must attract our sensory attention, and, since we rely most on vision to navigate our umwelt, most signs appeal to us by their visual presence.[6] When we notice something as a sign, we turn our gaze toward it and place it in our central visual field, but, being "significant," its function is to refer us to *something else*.

Now imagine that one significant object is a dark globular cloud approaching over the horizon. I first identify it as a thundercloud and then ponder its possible implications—e.g., heavy rain, flash flooding, lightning, brush fires, wind, falling trees, hail, etc. I have taken two rapidly sequenced cognitive steps: I know (1) what the dark globular object looks like (a thundercloud) and (2) what an approaching thundercloud may bring with it (the aforesaid effects). As Terrence Deacon (1997) has argued, the first step is based on my interpretation of the object as an iconic sign, the second on my interpretation of this icon as an indexical sign. These two online cognitive acts rely on my ability to access offline knowledge stored in semantic and episodic memory, but they are most immediately cued by the fact that I can process visual input using (1) my ventral visual stream to select the object from its ground and identify it and (2) my dorsal stream to calculate its spatial relation to myself and to other objects and, on that basis, assess its future implications.

So far I have been following Deacon's semiotic analysis and treating icons and indices as natural signs, made such by a perceiver's interpretation. But by doing so I have ignored the function of *intentional* signs, the fact that persons can, and regularly *do*, choose to share iconic and indexical information with on another. Other persons view the world from vantage points other than our own, so when we want the benefit of those extra pairs of eyes, we consider those signs for the knowledge they may provide. Being allocentrically situated sources of information, other persons serve somewhat as landmarks that also help orient us in space. Rather than plunging ahead, relying solely on egocentrically framed input, we stop, look for helpful sign-senders, get our bearings, and *then* move forward.

To return to my story of the thundercloud:

Having interpreted this phenomenon as an icon with serious indexical implications, I feel socially obliged to tell others about it, but, to accomplish this mission, I must *convert these natural signs into intentional signs*. So, when I reach the others, I will first need to attract their attention. That is, I will need to become myself a significant object to them, and they will need to recognize me as the person they have known me to be, i.e., a friendly, trustworthy bearer of information (step one), and then they must allow me-as-signifier to turn their attention from me to my message (step two). Having the ability to use conventionally coded sounds (symbolic signs), I might begin to share my concerns by saying something like: "Looks like we're in for a bad storm."

But suppose my companions and I did not share the same language or lacked vocal language altogether. Assuming they could not see the storm cloud from their vantage point, I might have to point skyward in that direction (an indexical sign), then cup my hands like cumuli and perhaps imitate the sounds of thunder (iconic signs). If we belonged to a species of prelinguistic, bipedal, highly social hunter-gatherers whose well-being depended on correctly interpreting weather signs as well as countless other natural phenomena, I would try to be as clear as possible when conveying indices and icons. And, if our species had had several million years to work out efficient means of communicating the knowledge necessary for it to have survived that long, I would probably have sufficient means to share my weather report.

Peircean semiotics can provide useful distinctions for researchers into the origins of language, yet few have chosen to use them. Terrence Deacon is one of the rare exceptions. As the title of his first book, *The Symbolic Species* (1997), implies, modern humans (*Homo sapiens sapiens*) became fully sapient only when they developed the capacity to convey, through the medium of speech, arbitrary, conventional signs, syntactically governed. He does not, however, consider the possibility that gesturally mediated indexical and iconic signs played any preadaptive role in this speciating breakthrough. Instead, he treats index and icon as though they were only natural signs that appear and are interpreted but are never consciously sent. As he characterizes them, both operate at lower levels of "referential competence," as mental processes that realize their communicative potential only in the symbolic code.

According to Deacon, the hierarchical structure crowned by speech has, at its lowest level, an iconic process in which one perceives an object, reacts to it on the basis of prior experience, and assumes that this percept corresponds to X or Y. This is what the vision researchers, cited in the last chapter, term object recognition, a match-up between a visually perceived object and the mental image type that corresponds to it.[7] As his example of an icon, Deacon describes a bird on a tree branch and a moth whose camouflage design allows it to melt into the bark pattern of the trunk: the bird looks toward the moth but assumes that what it sees is simply more bark and so misses his prey. The point of his example is that an iconic sign is an interpretive relation between a subject and an object in which the subject, by seeing *only likeness*, often fails to make proper distinctions. Iconic interpretation is equivalent to guesswork and, as a means of knowing the world, suffers from referential incompetence (Deacon, 1997:71–79).

Like icons, indices exist only as stimuli to be received and interpreted (Deacon, 1997:87). Weather signs, seasonal signs, animal tracks, physical symptoms, etc., are natural indexical signs interpreted as pointing toward a prior cause or a subsequent effect and are thereby temporally displaced from their meanings. Other unintentional indices are spatially displaced, e.g., the smoke from a fire or the sound that reaches our ears from an animal scurrying in the underbrush. As Deacon sees it, the advantage of indexical over iconic awareness is that it requires the observer to consider the spatial and causal relation of an object to its surroundings, rather than simply identifying it by its appearance.

Again, Deacon does not consider iconic and indexical signs as intentional signs that persons may use to communicate thoughts to others. This is neither an inadvertent omission nor, strictly speaking, a misreading of Peircean semiotics. In his famous definition of a sign, Peirce also omits mention of a sign-sender: "A sign . . . is something which stands to somebody for something in some respect or capacity" (1931:135). Deacon's use of Peirce is tactical and in itself well reasoned. I found it odd, though, that in a study of the evolution of language he was so quick to dismiss the possibility that iconic and indexical signs may have been precursors of symbolic signs. He is quite definite: the symbolic signs that make language possible are totally without precedent. There is a "lack of homology between language and nonhuman forms of communication. It is tempting to try to conceive of language as an interpolated extreme of something that other species produce, such as calls, grunts, gestures, or social grooming" (1997:34). Deacon is not in the least bit tempted to do so, even though those "other species" would have to include *Homo erectus* and its related species. Even Bickerton, who after all was no gradualist, assumed that gestures and vocal imitation preceded the so-called Big Bang, the genetic event that launched our own loquacious kind.

Could that first species of the human genus, *Homo habilis* (ca. 2.5 mya), have used a communicative code, some form of pre-language? We can be fairly certain that, at the onset of Donald's mimetic stage, humans began to transmit skills and behaviors, storing reusable routines in procedural memory. But what mediated this transmission? The ability to break a stone to make it a sharp tool could first have been learned by watching another do this, then by being directly instructed as to how the stone is chipped. But suppose, at some point in time, one person faces another and, using his two fists, pantomimes a particular technique of knapping, meaning: "Let's go out and make a blade we can use to scrape bark." Now there is more than the two-part learning experience of an observer imitating an action. Now there is an *addresser* communicating a *sign* to an *addressee*.

This last mimetic breakthrough, according to Jordan Zlatev, marked a transition from a simple transfer of information to a transfer in which "the subject intends the act to stand for some action, object or event for an addressee (and for the addressee to recognize this intention)" (2008:138). There is a three-stage progression: first, the ability to see and then do;

then, the ability to learn from hands-on instruction; and finally, the ability to perform an act intended to signify to an addressee some other act, object, or event. The latter semiotic act, which could easily form the basis of a gestural communicative code, is precisely what Deacon chose to ignore (ibid., 138; Frith, 2008).

Intentional visual signs must have been as important for the evolution of genus *Homo* as were symbolic auditory signs for subspecies *Homo sapiens sapiens*. Of course, many nonhuman animals can send and receive intentional signs, and not just in the visual modality. They can engage in multimodal, ritualized displays (aggressive and ludic), leave their scent to mark their territory, and produce location and alarm calls, all intended to communicate their default intention, which is, whatever the situation, to directly manipulate the behavior of others. Humans do this also, but, in addition, humans (and only humans) have learned to manipulate one another's behavior *indirectly* by sharing knowledge with them (Frith, 2008). This form of communication, which seems so natural to us, is alien to all other species. The thought that a conspecific might be interested in finding out this or that piece of information would never occur to a parrot, a dolphin, a horse, a bear, a dog, or even a bonobo. We humans not only have theory of mind and practice mind reading, but we have also devised a special way to assure ourselves that we can indeed read another's mind: we insert something into that mind via an intentional sign.

One way to perform that insertion is to point to something. Only humans use their arms, hands, and index finger to select an object with the intention of showing it to others.[8] This indexical act was undoubtedly the ur-gesture that accompanied our early hominid ancestors when, having descended from the trees, they stood looking over the high grass of the savannas, and began their journey across the vast surface of the globe. Long before iconic gesture, indexical pointing could have been a way of indicating "here," "there," and "yonder"; "forward" and "back"; "this" and "that"; and "I," "you," and "other."[9]

As a gesture, pointing is motorically sent, visually received, and intended to prompt the receiver to perform a particular visual action. Since the hominid visual system that early humans inherited was substantially the same as ours, and since pointing would activate that system, we can make several valid inferences concerning what visual signs could have been transmitted by this usually manual means. Human eyes, their irises framed with white scleras, would have also augmented manual pointing, especially when addresser and addressee were close enough. When the addresser's gaze was aligned with his or her arm, the addressee would be emphatically enjoined to look carefully. When the addresser's eyes were not aligned with the arm and hand, they might be fixed on the eyes of the addressee until those latter seemed aimed in the pointed direction. This would also be the case when the addresser (or "origo") used the hand, often the extended thumb, to point backwards.

Besides specifying the direction of vision, pointing can also specify what that visual action should be—e.g., whether the addressee should examine a narrow visual field, a broad field, or a randomly scanned field. Narrow pointing involves the extended index finger and specifies a narrow sector of the visual array. Broad field pointing uses the extended arm. This targeted field may be further enlarged by spreading the fingers and moving the hand in a continuous arc from one border of the selected field to its opposite border. Narrow and broad field pointing prompt central (foveal) and peripheral vision, respectively. An extended arm and a rapid movement of the hand and fingers over a broad targeted area would invite the addressee to perform a set of random saccades and fixations over that entire area.

Ever since hominids became bipedal, they would have communicated through such indexical signs, especially hand and arm pointing. This silent means of communication would have been especially useful in navigating unfamiliar and dangerous territory and in coordinating hunting. While we language users would regard their gestural repertoire as impoverished, it had to have served their purposes. Moreover, the gestures they used were probably more numerous and more specific than those I just described, e.g., hand gestures to indicate locations over-and-down, under-and-up, left-and-around, right-and-around, and so forth.

As intentional indices anchored in the present, these pointing gestures would have served as prelinguistic demonstratives. But where were the prelinguistic nouns and verbs? These, also anchored in the perceptual present, would have to be the perceived objects at rest or in motion that the pointing selected—the bounding hare, the nesting bird, the mist rising slowly from the valley. The prelinguistic earth was a world teeming with visible nouns and verbs. Early humans would thus have had all the indexical means necessary to share attention and coordinate responses to the details of this world.

Most of these gestural signs, being fitted to human physiology as it had evolved, would be universal within and across the various then-extant human species. Many others would be culturally transmitted within given communities and, being traditional, would tend to become conventionalized. Today, every culture has a set of indexical gestures, termed "emblems," some that it shares with other cultures, some that are uniquely its own. Emblems such as the hitchhiker's sidewise extended thumb, the approver's raised thumb, the rejecter's down-turned thumb, the index-and-thumb circle (short for "okay"), and the spread index and middle finger ("V," short for "victory") may be widely shared, but they are by no means universal. Possessing language, we do not have to rely on emblems to communicate, but prelinguistic humans may have maintained quite an extensive system of them in order to respond promptly to emergent circumstances.

What indexical gestures could not communicate, however, was information concerning objects and events that were temporally and spatially absent, e.g., the finding of a hive of honey *last year*, a reminder of the plum trees that grow on the *other side* of the mountain, or the promise to share a quantity of meat from *next month's* hunt. For this our speechless ancestors would need iconic gestures, manual gestures that traced the visual appearance or movement of some object, in order to evoke its image in the mind of an addressee. As I noted earlier, this kind of mental image is a set of features that form a simple prototype of an object. We call up that image when we encounter such an object in a visual array, but we can also use it to think about it in its absence. This significant adaptation, when it managed to link together multiple images in complex interactive scenarios, evolved into that powerful cognitive skill we refer to as "imagination." Once gestural icons were invented, probably at the peak of the mimetic stage (ca. 1.8 mya) with the appearance of *Homo erectus*, humans could for the first time invite one another to cross over with them into the realm of conceptual thought, a venture that could launch them *together* into the past, into the present elsewhere, or into the future—travels in time and space that, until then, no animal had ever undertaken.

Once iconic gesture came into general use in a given community, its forms would have been shaped by two opposed necessities: on the one hand, signs needed to be clear; on the other hand, they often needed to be rapidly transmitted. To serve the latter purpose, they would have to be abbreviated, but when doing so omitted significant distinctions, a potentially life-threatening uncertainty might result. Syntactical elements would have helped, but syntax evidently did not evolve until *Homo sapiens sapiens* appeared. So, when prelinguistic humans needed to communicate about objects and events not immediately perceivable and had only iconic signs to rely on, they were faced with a dilemma. As factors in information transfer, speed and clarity constitute a duality that may be aligned with parallel and serial processing, respectively, but an iconic sign cannot integrate both factors using a dyadic pattern. In other words, iconic gestures were capable of speed and clarity, but *not both at the same time.*

A possible solution to that problem might have been to use two styles of iconic signing. In situations in which speed was not needed (e.g., gossip and storytelling), signs would be fully shaped, augmented by pantomime and appropriate vocalizations. In urgent circumstances, provided these events were framed within clarifying contexts (e.g., a hunt or preparations for a winter shelter), signs could be delivered rapidly in conventionalized abbreviations. What I am suggesting here is that, before language emerged, iconic gesture could have been communicated both in "long form" and in "short form."[10]

Iconic gestures, in their capacity to represent absent objects, marked an advance beyond purely indexical gestures, but as a communicative medium they would be inherently problematic. Despite their limitations, however, iconic signs would have prepared humans to take the next step—to symbolic signs in the format, first, of a vocal protolanguage and then of a full language such as all living humans have inherited. For, like the indexical emblem, the abbreviated "short form" icon could also have become conventionalized to such an extent that it could function as a visual symbol and, rather than conveying the message "This resembles that," it would simply convey: "I mean that."

Hearing Voices

Before returning to our deep human past and exploring how and when voice became our preeminent means of mindsharing, I want to review some of what neuroscience now knows about the human voice—how it is produced and perceived and how vocal language is decoded by the brain. While we may never know exactly how and when we shifted from gesture-dependent communication to vocal speech, we can safely assume that we share most, though not all, of our own brain's sensorimotor circuitry with those ancestors of ours who underwent this semiotic transition. With that in mind, I will begin with some simple observations.

When contrasting gesture and voice, we need to consider not only the different way each is produced but also the different way each seems to be received. A visual object seems very much "out there," whereas a sound is a perturbation of the surrounding air that, when received, creates an effect that seems localized within our skull. The visual brain can usually locate a visible object with pinpoint accuracy, but the auditory brain can determine only the approximate source of a sound, and this by comparing the relative amplitude entering the right and left ears.

Other receptive differences demonstrate how these two independent perceptual systems complement one another. The fact that the eyes have lids that can shut off visual input allows us to sleep. While the visual system, turned inward, occupies the sleeping brain with soundless dream scenarios, the unshut ears and the ever-vigilant auditory system monitor the environing soundscape and, trained to discriminate danger cues from background noise, will wake the sleeping brain and its visual system only when particular sounds enter the ears. Urban apartment dwellers, for example, sleep peacefully through the wailing of police and ambulance sirens but wake up suddenly at a mere rustle at the door or on the fire escape outside their window.

The independence of hearing and vision also permits the waking brain to parallel-process bimodal information, e.g., to interpret a person's vocal tone while watching the expressive movements of his face, hands, and lower body. As in every multitasking action, however, our attention is not equally distributed: we choose information channeled through only one of these sensory modalities for focal attention, while absorbing information provided by the other modality using subsidiary attention—another instance of dyadic patterning. If, for example, the person before us seems to be nervously concealing something in his clothing, we will focus our attention on his hands and posture while monitoring any vocal sounds he makes for supplementary information, or, if we are conversing with someone, we may focus on her words while peripherally processing her body language, both intentional and unintentional.

Though gesture and voice, in reception, are processed by separate sensory systems, in production they are both motor actions. As Ulric Neisser phrased it, "[T]o speak is to make finely controlled movements in parts of your body, with the result that information about these movements is broadcast to the environment. For this reason the movements of speech are sometimes called articulatory gestures. A person who perceives speech, then, is picking up information about a certain class of real, physical, tangible ... events" (1976:156). While equally motoric in kind, the two differ in scale. Articulatory gestures operate at a speed and energy efficiency far greater than those of manual gestures. These tiny muscular movements are for the most part concealed within the mouth,

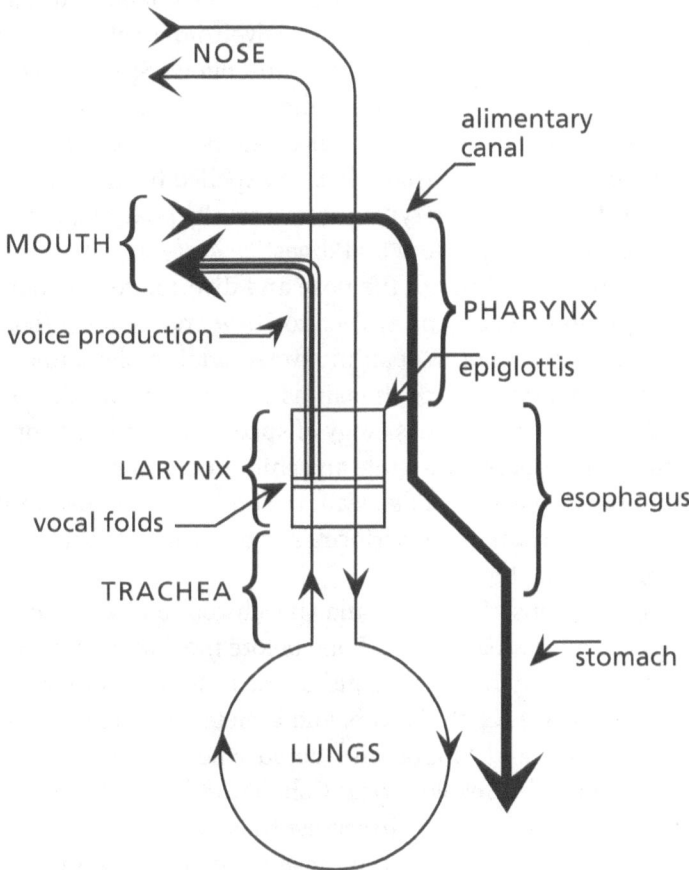

Figure 5.1 The physiology of voice production

where they recruit some of the mechanisms proper to the digestive system, but they are powered by voice-producing muscles in the chest and throat, recruited from the respiratory system (figure 5.1).

Speech, therefore, requires a precise coordination of vocal, as well as articulatory, muscles. Voice, which is necessary to make articulation audible, is produced during the exhalation phase of the breathing cycle when the column of air expelled by the lungs is first partially obstructed in the larynx by the vocal folds, or chords. Articulation occurs when the tongue shapes the oral cavity and directs the outflow of air past the teeth and lips. Long before speech evolved, primates had developed vocal strength and a degree of control through an exaptation[11] of the respiratory system. Vocalization (phonation) borrows the laryngeal valve, which (together with the epiglottal flap) serves to block food and liquids from entering the lungs, and converts this valve into a set of taut, vibrating folds capable of producing variations in pitch. Speech also requires articulation—the ability to produce distinctive phonemes out of which distinctive words are formed. For this an exaptation of oral anatomy was necessary: the column of carbon dioxide expelled from the lungs, having risen upward through the larynx, passes into the pharynx, the upper conduit shared by the respiratory and digestive systems, where it is turned from its normal exit through the nose and diverted to the mouth. Here, the teeth and tongue, organs evolved to chew and swallow food, process this now tone-laden column of air in reverse, adding phonemic features to it just before breathing forth this air as intelligible speech.[12] Even from this brief overview of the physiology of speech it should be apparent that fine-tuning this process was quite an achievement in multitasking. Since the sound-making aspects of speech, in addition to its meaning-making aspects, are important in the performance of verbal artifacts, I will return to this topic in chapter 7.

Our perceptions of gestures and speech sounds have also a common basis in the brain's motor cortex. Long before the discovery of mirror neurons, it was well understood that, just as the muscles of our arms and legs flex and relax as we watch dancers and athletes perform, the muscles of our chest, throat, and tongue also do so when we hear speech—even when we read words (Sokolov, 1972; Cohen, 1986). The fact that, when we hear speech, we concurrently innervate those muscles that we ourselves would be using if *we* were saying those words has prompted some psychologists to hypothesize that this inner mirroring of heard speech is not simply our way of intently following another's words but is in fact the principal way we have of decoding speech sounds and connecting them to form words. This is the "motor theory of speech perception," first proposed by Alvin Liberman in the 1950s. The major problem he addressed was this: our auditory system cannot detect the phonemes that form words unless each is distinctly uttered in an acoustic *series*, but the separate organs of speech often operate in *parallel*, thereby producing overlapping, co-articulated phonemes. Hearers of this acoustic stream can disentangle such phonemes only by mirroring the motor processes by which these sounds are articulated (Liberman and Mattingly, 1985; Galantucci et al., 2006).

Though the discovery of mirror neurons promised to give Liberman's theory a new lease on life, the unconvinced continued to explore the capacity of the brain to process purely acoustic signals. In the mid-

1990s a number of brain scientists, inspired by the findings of vision researchers, began turning their attention to the functional anatomy of hearing. Early research established that a ventral and a dorsal pathway proceeded from the auditory center (A1) of each hemisphere and conveyed information to the frontal cortex, where it was analyzed and appropriate responses were initiated. What was not yet determined was how that information was encoded and transmitted. By the end of the 1990s researchers were uncovering striking evidence that the functional anatomy of the auditory system, like that of the visual system, separately processes the "what" and the "where" (and "how") of the stimulus source (Rauschecker, 1998; Romanski et al., 1999). The ventral pathway, it was found, differentiates speech from environmental sounds, while the dorsal pathway calculates the spatial location of the sound source and its motion relative to the hearer. As with the visual system, the auditory paths each convey different information and do so in parallel one to the other, with some cross talk along their separate routes (Kaas and Hackett, 1999). Moreover, this double-routed system favored, as predicted, the left hemisphere for processing precise discriminations (Parker et al., 2005).

Just as Milner and Goodale (1995) had revised Ungerleider and Mishkin's (1982) visual model, finding that the dorsal path, or stream, processed not only the "where" of objects but, more importantly, the "how" of dealing with them through physical movement, Pascal Belin and Robert Zatorre proposed a revised model for the auditory dorsal path: "Anatomical segregation in dorsal and ventral auditory pathways reflects two different modes of auditory processing, analogous to those of the visual pathways. Applied to speech perception, our model suggests that the dorsal pathway extracts the verbal message contained in a spoken sentence, while the ventral pathway is responsible for identifying the speaker" (2000:965). This revised model implies that, like its visual counterpart, the auditory dorsal pathway has a greater access to working memory and can process larger chunks of information, i.e., multiple phonemes as they form phrases and sentences. "Similarly," they continued, "the dorsal pathway processes the melody of an instrumental piece, while the ventral pathway recognizes the instrument by its timbre" (966).

These findings hold important implications for the poetics of oral performance. This dorsal pathway appears to be the brain's means of actively following serial auditory events. Hearing a story is therefore analogous to moving through objects in a spatial field, but now, instead of a landscape, we are moving through a soundscape and, instead of an optic flow, we are processing an acoustic flow. The analogous function of these two dorsal pathways, the visual and the auditory, finds expression in a number of common idioms: "Let's explore this topic," "where is this leading," "I'm following the story," "I see where you're going now," "I grasp what you're getting at." In subsequent chapters I will examine the Greek term for narration, *diêgêsis*, and the metaphorical significance of its root meaning, "leading through."

The research of Greg Hickok and David Poeppel (2004) suggest yet another interpretation of how the auditory pathways process speech. Finding Milner and Goodale's model even more closely applicable to speech perception, they report that the "ventral stream . . . is involved in mapping sound onto meaning," while the "dorsal stream is involved in mapping sounds onto articulatory-based representations" (2004:67). Like its visual counterpart, the auditory ventral stream would then be able to access semantic memory—now "semantic" in the specifically linguistic sense of the word. The dorsal stream, like *its* visual counterpart, would be connected with motor areas—in this case, areas that control the organs of speech production. The perception of speech would therefore involve two parallel-operated processors: one that identifies the source and analyzes the meanings of speech sounds (the "what"), and the other that monitors the articulatory gestures that produce them (the "how"). If Hickok and Poeppel prove correct and these two pathways are in these ways complementary, then the claims made by the advocates of the motor theory of speech perception and the more mainstream claims of those who hold that speech sounds are processed as purely acoustic phenomena may someday be reconciled.

Much of what contemporary neuroscience has revealed concerning voice and hearing and their working relationship with visual perception would also be applicable to prelinguistic humans. If, as many suppose, visual gestures provided early humans with the means to communicate their knowledge, wishes, and plans, we may suppose that vocal sounds were also significant. An obvious example would be iconic sounds, such as animal imitations. But there were no doubt other uses. As involuntary indices of mood and emotion, vocalizations such as laughter, sobbing, groans, shrieks, growls, and sighs would, of course, be immediately interpretable alongside visual signs, such as intentional manual icons and unintentional orofacial and postural indices. Like other spontaneous expressions, these vocalizations too could be ritualized and conventionalized (cf. forms of laughter used to indicate embarrassment, scorn, friendliness, etc.).

Vocal signs could also be directly incorporated into gestural discourse. One reason to do so would be to attract the attention of a set of spectators. Just as raising one's hand is now a way to get others to listen to one's voice, raising one's voice would have been a way to get others to look at one's hands. Then, during the unfolding of the gestural message, the sounds associated with particular affects could be modulated as to pitch and amplitude and synchronized with the gestures. For example, high and loud vocalization might be associated with urgency or rising passion; high and soft, with fear; low and loud, with threatening demeanor; and low and soft, with gentleness or comfort. Gradually rising or descending pitch might represent intensification or relaxation or an upward or downward movement in space. Finally, voice might have been used to "punctuate" gestural discourse. As David McNeill (1992) has observed, speakers regularly use certain downward manual chops, or "beats," to mark important words or transition points.[13] Perhaps vocal "beats" were similarly used to emphasize particular handshapes and partition gestural sequences.

What I am suggesting here is that if gestural communication was regularly accompanied by vocalization, we might call the latter "paragesture" on analogy with those phenomena we call "paralanguage," specifically those gestural signs that now accompany speech. In either case, while one of the two communicative channels would normally receive focal attention and the other would receive subsidiary attention, both would be processed in parallel and their information correlated as an audiovisual dyad.[14]

As I proposed earlier, the need to communicate quickly, accurately, and in all circumstances must have applied continuous pressure on humans to find the means to do so. While indexical gesture had been an early means of pointing others' eyes toward objects in the immediate here-and-now, iconic gestures, either fully signed or condensed into emblems, could have been used to point others' minds to objects beyond the here and now. But this iconic code would have had its inherent limitations: it could not be used when the hands were occupied with tool making or tool use, or with gathering or transporting objects. Furthermore, since it depended on sharp visual perception, it could not be used in dim light or darkness, in dense foliage, by persons not facing one another, or over broad distances. Despite such limitations, however, gesture must have proven superior to voice as a medium of communication for well over a million years. Though voice was useful as an index of mood and emotion, as an *iconic* medium for object information, it was limited to representing things that made sounds, e.g., birds, beasts, thunder, and wind.

These were some of the factors that inhibited pre-sapient communication, but equally important to consider were the interim solutions that may have been adopted. The development of abbreviated forms of iconic gesture would have introduced conventionalized (arbitrary, symbolic) elements into what remained a basically iconic code. This might have entailed the use of "determiners," simple gestures attached to air-pictures that might designate pronominal persons, tenses, and the like, and may also have recruited conventionalized (nonindexical, noniconic) vocal sounds for the same purpose (Arbib, 2009b).

Protolanguage, the Long Transition

That old problem facing advocates of the gestural origin of language has been to account for the transition from gesture to voice. This becomes less of a problem when we recall that nonhuman higher primates regularly use gesture *and* voice to communicate their wishes and emotions. While human primates must have been at first better at transmitting information through gesture than through voice, at some point voice became the preferred vehicle. But what if this "point" was a transitional period of over half a million years, say, from the appearance of *Homo erectus* to that of archaic *Homo sapiens*? And what if, during all this time, humans regularly communicated bimodally, only gradually shifting from a code that foregrounded gesture to one that foregrounded voice, the latter becoming the fully dominant vehicle only with *Homo sapiens sapiens* (ca. 150,000–100,000 B.P.; McNeill et al., 2010)?

The hostile environment that confronted, first, bipedal hominids and, later, tool-making humans amply supplied selective pressures to which they responded by fashioning further techniques. As I have proposed, the collective need to respond quickly to emergencies favored the invention of abbreviated gestural signs (emblems), a "short form" semaphore system that sacrificed iconicity for speed. Over time these would have become fully symbolic signs—conventional, arbitrary, and, as Saussure termed them, unmotivated. What each of these emergency signs meant was what the group had already agreed it meant. When an emergency arose, a leader might vocally convene the group to transmit to them a rapidly gestured series of visual symbols. But this whole procedure could be further accelerated by assigning symbolic meanings to its *vocal* announcement, which would now combine two sign functions, the attention-grabbing alarm cry and a portion of the message that would normally follow it. A gradual transition, such as this, from (1) predominantly iconic gesture to (2) predominantly symbolic gesture, then from predominantly symbolic gesture to (3) predominantly symbolic vocalization would constitute a natural three-step progression that would mark the long transition of protolanguage, a process through which the language-ready brain was organized. The formulaic theory, as proposed by Alison Wray and subsequently supported by Steven Mithen (2006) and Michael Arbib (2009b), seems to me to fill the gap between a gestural code and full language far better than Bickerton's theory of a grammarless pidgin (Wray, 1998:47–49).

But what of gossip, the social information activity that Dunbar named as the replacement for primate grooming? To engage in this form of alliance building one would have to go beyond the list of generic greetings, requests, and commands. One would have to cite specific persons, to "name names," and mention specific objects, e.g., things that X took from Y or refused to give to Z (Wray, 1998:50). In short, one would have to be able to utter single words—proper and common nouns. The function of common nouns could still have been supplied by iconic gestures, a communication skill that, until full language appeared, would have been available for exchanging object information. But a proper noun, in the form of a separate manual icon reserved to designate an absent *person*, one of perhaps over a hundred members of the community, would seem awkward to archive, whereas a vocal label would be relatively easy.

Would gossip as a mechanism of social regulation have to wait until fully language-capable *Homo sapiens sapiens* evolved? My guess is that, if premodern humans could indeed communicate through a holistic vocal code during that span of perhaps a million years, they could also find a vocal way to refer to absent persons and talk about them behind their backs. After all, a proper noun, in this case a personal name, is a unique item of social information and, as such, would no more require a supporting syntax than would a pointing finger. It seems likely, therefore, that separate vocalizations of personal names would have been increasingly used to supplement a holistic, formulaic code during the million years when *Homo erectus* was the dominant human species.[15]

The timeline/flow diagram of figure 5.2 is divided vertically into visual kinesic signs and auditory vocal signs. To the left (visual) and right (auditory) extremes lie unintentional indices, clues that humans

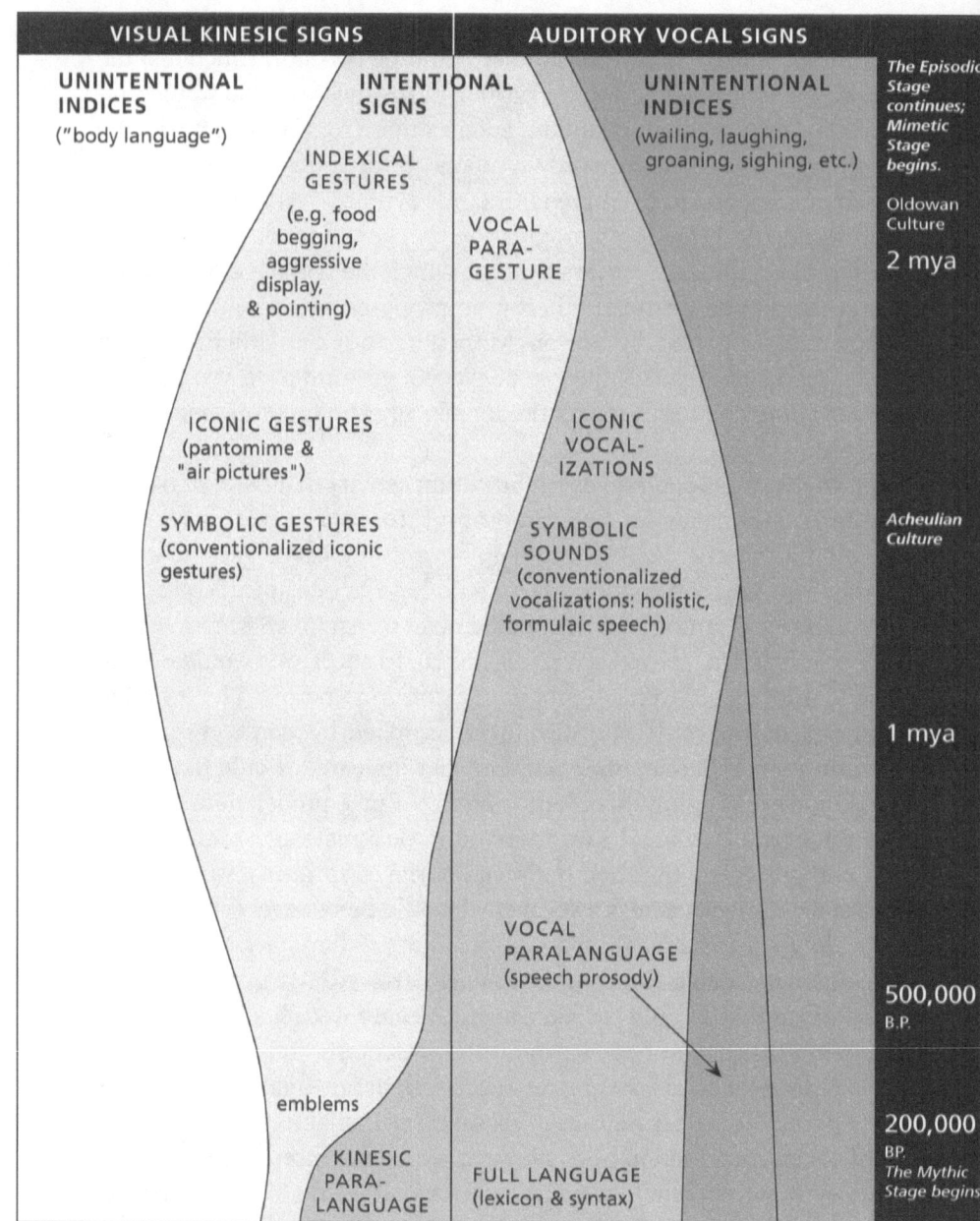

Figure 5.2 The evolution of semiotic skills from gesture to speech

spontaneously reveal about their thoughts and emotions. In the middle section lie intentional signs, visual and auditory information that they deliberately communicate to others. As the diagram indicates, human communication has always been a variously proportioned hybrid of seen gestures and heard vocalizations. Since they are produced and perceived independently from one another, gesture and voice can operate in parallel, each complementing the other. As we have seen, when separate complementary functions operate together, they conform to the dyadic pattern, one requiring narrow focal attention, the other, broad awareness, one centralized as figure, the other peripheralized as ground.

Assuming that gesture was the earliest medium of human communication, it would have been complemented by what I have termed vocal "paragesture." By enclosing them in the same shade of gray I mean to suggest that gesture and vocal paragesture were integrated in a bimodal communicative code, one in which gesture received focal attention, while its vocal accompaniment received subsidiary attention. At first these sounds must have resembled the barks and hoots of latter-day chimps and bonobos. Only gradually did the human voice attain the articulatory control necessary, first, to mimic environmental sounds, such as animal and bird calls ("iconic vocalizations"), and later, to utter a range of phonemes that could be serially linked and conventionally assigned particular meanings ("symbolic sounds" and "full language"). The transition from a predominantly gestural code to a predominantly vocal one was, as the diagram indicates, a slow process, and, during the long heyday of *H. erectus* and related species, gesture and voice would have shared the tasks of communication. But, for reasons that I summarized in the last section, the vocal medium proved more successful and, as a basis for symbolic signs, came to translate the human umwelt into a virtual universe of discourse.

When full language emerged (ca. 150,000–100,000 B.P.), it came to dominate human communication, but it did not wholly eliminate the older semiotic media. The oldest of these, the unintentional vocal indices, it converted into a vocal paralanguage of controlled, intentional, prosodic features. The next oldest communicative medium—indexical, iconic, and symbolic gestures—it converted into a kinesic paralanguage mainly of hand and arm movements dedicated to emphasizing and punctuating verbal discourse. Emblem (i.e., symbolic gesture), the last surviving element of the once-dominant gestural code, was an element that language did *not* integrate into its own symbolic sign system. Instead, emblem remained as a potential medium for ad hoc gestural communication among persons lacking a common language and, by assimilating lexical and syntactic features, has provided the basis for fully developed signed languages, which in some respects resemble the hypothetical model I proposed earlier, the iconic "short form."

This diagram also illustrates another point: those features of the once-dominant modes of communication that became expressive accompaniments to speech also became the identifying features of oral poetry. In my final chapter (chap. 7), I will propose, among other things, that prosodic structure and performance techniques are stylizations of vocal and kinesic paralanguage and that formulaic diction is inherited from holistic protolanguage. As I suggested at the beginning of chapter 1, imaginative writing, being a living link to our phylogenetic past, derives its special properties from its power to actualize those older, deeper cognitive levels that still remain within us.

six
Language
ITS PRELINGUISTIC INHERITANCE

The cognitive approach to language accepts the old premise, at least as old as Aristotle, that words and sentences are the expressions of actions, conscious and unconscious, occurring within the mind. Rather than considering language an autonomous system with a complex external set of slots for human meaning, cognitive linguists view it as a means by which internal meaning is discovered, organized, stored, and shared. They have therefore sought ways to explore the workings of the brain/mind that lie beneath language, which language seems designed to reflect—e.g., memory and its various systems, perception and its several modalities, and the figure–ground pattern used to organize experience (Fauconnier, 1999:96, 97).

Borrowing a term from Erving Goffman, in his *Presentation of Self in Everyday Life* (1959), Gilles Fauconnier called these various elements "backstage cognition." Having investigated the contents of this "backstage" in the preceding chapters, I now come to a crucial turning point in this story, the advent of Donald's mythic stage when full language (lexicon + syntax) emerges and profoundly alters the human mind, not so much by changing it, but rather by becoming its mirror wherein it can observe its own deeply established workings. The cognitive linguistic issues I will now raise reintroduce some of the cognitive evolutionary topics I have already examined, such as tool use, play, visual perception, and pre-grammatical communication. These functions, having helped shape the language-ready brain, continue to operate below its surface and, as we shall see, the dyadic pattern that informs so many of them reappears in the general structures of grammar.

The Rhetorical Motive

Getting others to "take our word" for something is our principal motivation for speaking. Yet getting others to exchange their own sense of reality for these words of ours is not always an easy sell. We must be believable. Our intentions must be, or seem, transparent. Our information must not only be ours but the rightful property of all who will accept it. For all this there is but one skill and that skill we call "rhetoric."

Language is essentially rhetorical—deeply rhetorical and intensely rhetorical. Did that sentence sound rhetorical? I suppose so. If I had said, "Language, I submit, is a medium by which speakers attempt to persuade others of the rightness or wrongness of a given course of action," I would have sounded less vehemently rhetorical and, to some perhaps, more persuasive. But that's just the point, whatever way I or you or others make a statement—even if what we say is "I have no interest in persuading you"—we do so to persuade others that we are right.

This has been going on for a very long time. Even before spoken rhetoric, the rhetorical motive existed as persuasive tactics, e.g., displays used to establish or maintain territorial feeding and breeding rights or attract mates. Whether communicated by sound, gesture, or smell, these messages conveyed information as to the relation of the sender to prospective receiver(s), information as to ranking (dominance, submission), need, danger, location, alliance building, sexual interests, etc. Among conspecifics this is social information, and among highly social animals (e.g., primates), its exchange takes on greater importance and complexity (Kennedy, 1992).

Our primate cousins, the chimps and bonobos, spend a considerable portion of their waking time in persuasive, or, as we call it, "manipulative," behavior. Grooming, begging, menacing, sharing—these behaviors help maintain social cohesion and, we may suppose, they did so for our last common ancestor some 7 or 6 mya. This exchange of social information long preceded that of object information, and, when the latter appeared as gestural signs at some point prior to the first stone technology (2.5 mya), it must have been subsumed within the functions of social information. Object information must have been an advantageous possession, for the more one knew, or could persuade others that one knew, the higher social status one could enjoy (Bax, 2009).

When it came along, spoken language must have greatly enhanced the prestige of those who had amassed object information. As we know, some speech is still power and there are some speakers before whom an audience remains silent. This resembles the rhetorical primacy of the alpha male in a wolf pack or the silverback in a band of gorillas. Yet the power of speech is not absolute. We normally play two roles, the speaking "I" and the hearing "You." Practiced at a low level of rhetorical intensity, this turn-taking most typically characterizes everyday domestic conversation, the speech of intimates. In a relation of mutual trust, credence is implicit. Conversants do not need to be persuaded in situations that are wholly familiar, such as "Here's food. Sit down and eat"—"Thanks" or, as Wray and Grace (2007) might phrase that in "esoteric holistic language": "'Sfood. Sdown'neat"—"'anx." In such exchanges the division between "I" and "You" is not sharply delineated, and the two seem connected by the inclusive first person plural, the "We." The separation between the two pronominal roles would be more definite with the "exoteric" code when, in "speaking with strangers," issues of hospitality and gratitude need rhetorical expression.

Speech as an instrument of social intelligence is strongly implicated in the formation of moral structures. Every "I" indexes a spatiotemporal center, a "here" and "now." But, as Mühlhäusler and Harré (1990) point out, every "I" has a "double indexicality": it locates the "I" not only in physical geography but also in "moral geography." As a physical index, the first person pronoun "tends to be universal and stable and should be registered in all languages" (95). As a moral index, it refers to the "person who is to be held morally responsible for its illocutionary force and its perlocutionary effects" (92).

One motive that drove the evolution of language, according to Terrence Deacon, was the need to gauge the "future probability of [others'] altruism on the basis of past experiences"(1997:398). This was a particularly crucial element in sexual selection: the hunter who shared his kill with his mate was permitted sexual access. "The pair-bonding relationship in the human lineage is essentially a promise, or rather a set of promises that must be made public. These not only determine what behaviors are probable in the future, but, more important, they implicitly determine which future behaviors are allowed and not allowed; that is, which are defined as cheating and may result in retaliation" (ibid., 399).

If symbolic signing, gestural or vocal, had been a necessary medium for promises, it had to have co-evolved with tool use, because it was mainly their mastery of tools that provided humans something worth promising to one another. Like a tool, language also extended the power of its user—in this case, the power to negotiate social contracts. At least since the inception of full language, most humans ever born would have been offspring of promise-bonded parents, and so the traits associated with reciprocal altruism would likely survive to influence the emergence of related traits.

As Deacon sees it, symbolic language and sex-for-food bonding co-evolved. "Symbolic culture was a response to a reproductive problem that only symbols could solve: the imperative of representing a social contract" (ibid., 401). Since a contract obliges the parties who enter into it to perform stipulated actions, some of them of a very general sort, representing these obligations using an old-style semiotics would pose considerable problems. Knight, Studdert-Kennedy, and Hurford offer this appreciative summary of his theory:

> Deacon's insight was that nonhuman primates are under no pressure to evolve symbolic communication because they never have to confront the problem of social contracts. As long as communication concerns only current, perceptible reality, a signaller can always display or draw attention to some feature as an index or likeness of the intended referent. But once evolving humans had begun to establish contracts, reliance on indices and resemblances no longer sufficed. Where in the physical world is a 'promise'? What does such a thing look like? Where is the evidence that it exists at all? Since it exists only for those who believe in it, there is no alternative but to settle on a conventionally agreed symbol. (2000:9–10)

A social contract, then, is an abstraction projected into a new conceptual domain, the human future. It is a conventionally agreed upon arrangement that can only be represented by a conventionally agreed upon set of signs. Words, as vehicles of the contract, can represent that contract because each one of these *is itself a contract*. I can "give my word" to you because it is not only *my* word but *our* word, a symbol the significance of which you and I both agree on, an a priori agreement upon which every subsequent agreement is premised. Consequently, this "symbolic species" of ours, after millions of years trying to mind-read one another's intentions, now left its hominid cousins behind in the bush and went off to settle an unseen savannah of discourse that can never fully exist, a future that only promises to be real.

In the pronoun paradigm, widely regarded as a linguistic universal, language has preserved a standard social relationship. The Greek grammarians referred to it in theatrical terms: pronouns were *prosôpa*, "face masks" (Latin, *personae*, i.e., devices *through* which the voice *sounds*). If one were to elaborate this analogy and try to visualize the "I" mask, it might be mainly mouth, the "You" mask mainly ears. The third persons, the "They," as perceived from the vantage point of the "I" and the "You," would be somewhere else, out of earshot or otherwise excluded from the speech event, and might therefore be represented by an earless, closed mouth mask.

In the actual speech event, the "I," as long as he or she holds forth, occupies the center of the hearer's attention, and, if more than one hearer is in attendance, the audience forms a circle about the speaker. Outside this acoustically determined circle lie the others who do not hear or do not heed the *I*-generated speech. This separation of the third person from the locus of speech has led some linguists to deny outright the status of persons to *he, she, it,* and *they* (Benveniste, 1966/1973:197–98). As conscious subjects, they have their own pronominal relations (theory of mind accords them that), but from the egocentric vantage point of the centered *I* and the encircling *You*, the *They* are nominal particulars. Until contact is made with them, they remain the foreigners who "don't speak our language," the enemies to whom "force is the only language they really understand," the heathen who "need the word preached to them," or some other noun designation.

In his study of social control, *Power, Influence, and Authority: An Essay in Political Linguistics* (1975), David Bell asserts that "politics is talk"—more precisely, "*who talks to whom, when, how*" (10, author's emphasis). Though he does not correlate this talk with the pronominal paradigm, his three degrees of social control may, with a little rearranging, fit neatly into it. Citing his title, suppose we begin with *influence*: "If you do X, you will do (feel, experience, etc.) Y." The relation between the speaker and the hearer in this case is one of adviser. A consequence will occur that *I* will not produce: *You* will bring it on yourself. This is the speech produced within a dialogic *We*, an inclusive entity that shares information and confers on actions to be taken. In this typically shifting exchange between relative peers, each *I* seeks to influence the group. The most influential *I* will tend to lead the group toward some particular resolution of its perceived problem. What Bell calls *authority* relies on automatic compliance with a communication. This *I* never argues a course of action, never descends to hypotheticals. The commanded, whom I have referred to as the absolute *You*, is one who has "internalized values or norms favoring obedience to such communications from a particular source on a particular range of subject matters" (78). It is only when the authority of the commander fails that hypotheticals are once again resorted to: "If *You* do X, *I* will do Y." Note how this differs from "influence." *I* is the unambiguous agent in what Bell calls "power speech," and Y is either a threat or a promise or both. "Power implies the existence of a valued object that (a) can be manipulated (i.e., increased or diminished by one actor with respect to another), (b) is valued by the respondent, (c) is in relatively short supply, and (d) is divisible. Any object fulfilling these criteria can become the basis of a power relationship" (82–83). The actual execution of a threat by an *I* upon a *You* is not a speech situation, for at this point the *You* is neither to be influenced nor commanded through speech. This "you" is now no longer a second but a third person and must now either be bought off with a gift or vanquished by brute force.

The second person is thus precariously located between the first and third persons, between the status of a would-be member of a *We* susceptible to influence and that of a banished *They* subject to acts of power. The *I* always tacitly judges the *You* as either "one of us" or "one of them," meriting the presence of the *I* or consignable to the outer darkness of the *They*. Hearers as members of the *You* are never explicitly aggregated to the *You*, as in the statement "You are one of you," because *You* has no capacity to define itself as an independent group, no internal solidarity. Its ultimate aspiration is always membership in the speech of the authoritative first person(s). Its ultimate fear is reduction to the nonspeaking, nonhearing status of the third person, the status of a virtual "nonperson."

Language Play

Whether symbols first began as conventionalized gestures or vocalizations, they took over the work of indexical pointing and iconic handshapes, pantomimes, and vocalizations. For immediate social information, references to and perceptions of the *I* and the *You* were always sufficient, but once perceptually absent third persons could be signified, social information was massively enhanced by the inclusion within it of objectifiable others and their behaviors. Language permitted talking behind third persons' backs even when those backs were not there to be pointed at.

As I have suggested in the last chapter, there would have been other persistent selective pressures on humans to develop faster and more accurate means of communicating social and object information. During a period of perhaps a million years (the so-called Oldowan culture of *H. habilis* and *H. ergaster*), while conventionalized manual iconic gestures would have been the most efficient means of sending signs, the need to coordinate collective responses to sudden emergencies would have favored the invention of context-specific, abbreviated manual signs, in effect, arbitrary gestural symbols, or emblems. Vocal indices, at first used in order to draw attention to the gesturer, would have later become vocal symbols once a system of holistic utterances was developed (ca. 1.5 mya, the beginnings of the Acheulian culture of *H. erectus* and related species). Augmented by gesture, a holophrastic system would then be maintained for at least another million years. As Alison Wray has proposed, separate protolanguages of this sort would have been transmitted within isolated communities until "talking to strangers" necessitated the segmentation of social formulas into syntax-governed words for the main purpose of exchanging object information. If a community of *H. erectus* could manage their affairs with a lexicon of, say, a hundred holistic vocal symbols, a community of *H. sapiens* might now need a lexicon of over a thousand individual symbols.

Leaving aside for now the issue of syntax, this lexical explosion was made possible by what Charles F. Hockett in 1960 named "duality of patterning," the last of thirteen "design-features" used in animal communication, only the final four of which are unique to human animals. "The meaningful elements in any language—'words' in everyday parlance, 'morphemes' to the linguist—constitute an enormous stock. Yet they are represented by small arrangements of a relatively very small stock of distinguishable sounds which are in themselves wholly meaningless" (6). This ability to use strings of meaningless phonemes to generate a humanly inexhaustible number of meaningful morphemes ensured that every language could have a vocabulary adequate to all its speakers' conceivable needs.[1]

Hockett's duality of patterning explains how language generates vocabulary, but not how humans manage to accomplish this alchemical feat of transforming a series of wholly meaningless sounds into a series of fully meaningful representations. Genetics may eventually prove that

LANGUAGE: ITS PRELINGUISTIC INHERITANCE

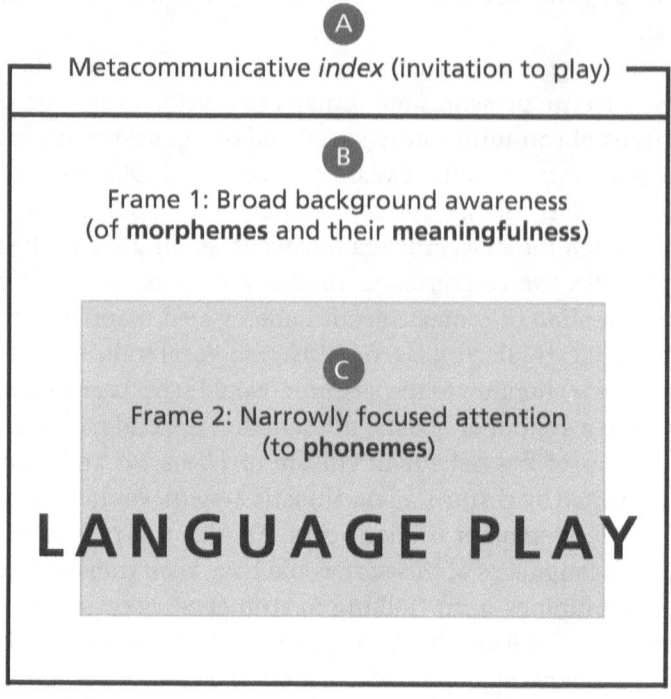

Figure 6.1 Speech comprehension as dyadic play framing

a random mutation (e.g., the FOXP2 gene, ca. 200,000 B.P.) rewired the pre-sapient brain in such a way that the appropriate voice, hearing, memory, and articulatory centers became coordinated, but, if this did indeed happen, it succeeded only because preadaptive mechanisms had already made the brain language-ready.

One of these mechanisms, I would submit, may have been the primate play instinct, expressed specifically as *social* play (see chap. 3, pp. 62–68). This play form, as adapted to human communication, I will simply call *language play* (figure 6.1).[2] This play begins with a metacommunication, an index that could be a gesture or vocalization used to draw attention to a potential speaker: I could, for example, raise my hand or clear my throat, or I could simply articulate the first one or two phonemes of the utterance. This latter sign, being acoustically distinct from any involuntary sound, such as a yawn, a laugh, a groan, or a hiccup, would signal that a series of words are on their way.

However I produce it, this metacommunication is the "Let's pretend" cue that invites all within earshot to play with me. This means that, if one attends focally to the whole series of phonemes I am about to utter (C), one will find that that they resolve themselves into a series of morphemes (B) that will constitute my message. The relation between C and B is wholly conventional. The subsequent relation of B, the string of words, to the meaning they produce is equally conventional, but Hockett does not choose to include this latter relation among his "design-features." His focus remains on this one, crucial semiotic innovation. In the beginning was indeed the word, or, as he preferred to call it, the morpheme. The sentence and the syntax that governs it would follow the same basic principle, but if we interpret "duality of patterning" as Hockett intended, we will appreciate it as the key that opens the universe of discourse.

Much attention has been given to the anatomical changes in the vocal tract and the fine-tuning of the brain's motor areas that control the breathing and tongue muscles. Much evidence has also pointed to the social pressures that reward improved communicative skills. These lines of research are certainly valid ways to understand the preadaptive traits underlying the emergence of language. There remains, however, that other very big question: How does a child master words when the relation between sound and meaning is wholly arbitrary? Noam Chomsky (1980), who continued to oppose the behaviorist claim that language is first learned by a process of stimulus and response, argued that the amount of time used by parents to teach this behavior could not account for the complex grammatical learning that the child achieves by three years of age. His "poverty of the stimulus" argument reinforced his theory that grammar at a deep structural level is an innate property of each member of our species and that each of us is born with a Language Acquisition Device (LAD), a specific brain module wholly dedicated to language.

Children's aptitude for language does seem extraordinary. We all know how difficult it is to acquire a second language in adulthood, whereas small children, even before they can hold a glass and drink from it, even before they are toilet-trained, can manipulate a sizable lexicon and a fundamental syntax. Developmental psychologists have listed a progressive series of stages, each leading on to the next, from birth to two years and beyond. In a paper that strongly influenced Alan Leslie's seminal study of play behavior (1987), Lorraine McCune-Nicolich (1981) correlated early pretend play with language development. Between 9 and 24 months, children engage in pretend play and begin to utter referential words (i.e., single nouns that designate third-person objects), two activities that "indicate the beginning of a separation between the means of signifying a meaning and the meaning itself. In play, the child shows awareness of a meaningful action (eat, sleep) and the potential for using that action in play (e.g., pretend eat, pretend sleep). . . . In language, the child shows awareness that words are meaningful and that such meanings can be used for communication" (792). Describing a further developmental stage (24–30 months), she noted that "[b]oth language and [pretend] play proceed from single units to combinations. It may be that a shift in cognitive functioning, such that [linguistic and play] units can be combined, underlies the development of sequential behaviors in both domains" (794).

The basic play principle of double framing (p. 67) seems to apply to the child's acquisition of the lexical components of a symbolic code. The word "dog(gie)" is itself not that furry barker with the wagging tail—we just pretend it is in order to talk about that animal. The banana that becomes the telephone is not actually a telephone, and, by the same measure, the word "banana" does not have a yellow peel and a soft, sweet interior. What, better than language, appeals to the playful mind of a two- or three-year-old? Words do not at all share in the sensory features of the world they signify, so learning these arbitrary, "made-up" signs, these creations of non-sense, becomes child's play. The child, in effect, reasons: "If you want to say that that is a 'dog' or a 'banana,' whatever, I'll play along . . . Now give me another word to play with." Of course, the adults are usually delighted by this performance and do their part to reinforce it.

Small children take great pleasure in learning to control their motor systems, first crawling, then standing, walking, and finally running. They also enjoy learning how to use their hands to grasp and manipulate objects. Part of the satisfaction they feel in language acquisition, an additional payoff for their attentional efforts, is undoubtedly their sense of controlling *mental* objects. Language play recapitulates the evolution of the pronoun paradigm, which, as I proposed earlier, progresses from a simple social information code, the *I/You* relation, to one that includes within in it a capacity for third-person object information. As the initial stage of language acquisition, children's language play is closely associated with object permanence, the realization that objects, like words, are context independent and can appear in any number of surprising settings (Lifter and Bloom, 1989). It is also associated with the early object play that Winnicott (1982/2005) examined, that projection of second-personhood onto dolls, stuffed animals, and all sorts of manipulable items that occupy the space between the child's ego and the principal caregiver, typically the mother. As Winnicott speculated, this object play prepares the child to accept the adult projections we know as the visual arts of painting, sculpture, and architecture (Rudnytsky, 1993). What I am proposing is that language play, as the manipulation and projection of verbal "objects," is the basis of storytelling, fictionality, and the making of verbal artifacts.

As adults, we have forgotten the fun of early childhood language play, our first realization that words are magical. As modern humans we have also lost that sense of wonderment that our first symbol-exchanging ancestors must have felt. From time to time, though, we are reminded of this play when a particular phrase becomes salient for its phonological features. Phonological play occurs when the phonemic level becomes foregrounded and the morphemic level recedes, as when the same phonemes recur in a passage with more than normal frequency, either as consonant sounds (alliteration) or vowel sounds (assonance). Symbolic signs become opaque and cease to operate solely as signs. This shift from sense to sound is rarely complete, but the pleasurable effect it sometimes has suggests the reawakening of old vocal behaviors, such as primate vocalization and infant babbling.

There is yet another kind of language play that foregrounds the semantic content of a word. It does this not to draw attention to itself so much as to shift attention *from* it to some other word or referent. This play, which has its roots in gestural modes of communication, we might call deictic play in that it points to or represents some other imagined entity. In speech or literature we are invited to engage in deictic play whenever we encounter a particularly apt metonym or metaphor.[3]

The importance of these two tropes may stem from the fact that they represent the superseded sign functions of index and icon. Metonymy prompts the mind to locate a target by using a metonym, usually a concrete noun, as a reference point. In the saying "The pen is mightier than the sword," each noun is a tool that points away from itself toward its typical users and their typical activities. This indexical relationship replicates in words the use of landmarks in mental mapping and wayfinding, activities connoted by the term "diegesis," literally a "leading through." A metaphor image, when introduced by a speaker, prompts the mind to relate it to another image or concept, implied or stated, that is stored in a different semantic domain. It is to this, the target of interest, that the metaphor image imports from its source domain particular properties that may be applied to this target. Thus, when Romeo exclaims, "Juliet is the sun!" his focus is on Juliet, while the sun is a mental entity that happens to share certain iconic properties with her, e.g., light, warmth, centrality, and the power to awaken, but not rotundity, the power to blind, or the tendency to dry crops during a drought. Though metaphor, as a verbal form of icon, is a *jeu d'esprit*, a mind play, it has an outward analogue in *mimesis*, imitative pretend play. In a mimetic performance, as in a verbal metaphor, two entities, actor and role, are simultaneously equated and differentiated. When symbolic signs step forward to perform the roles of indexical and iconic images, the play principle that is preadaptive to language but normally hidden in "backstage cognition" becomes suddenly illuminated and prompts us, as Emerson said, to "clap our hands in infantine joy and amazement."

From a phylogenetic perspective, language play may be viewed as a linguistic rhetorization of older communicative traits associated with System-1 features. Yet like most of those uniquely human traits, language, the quintessential S2 cognitive activity, incorporated the older S1 codes it superseded, doing so by converting indexical and iconic signs (predominantly visual) into symbolic signs. The emergence of language proved perhaps less of a shock to the system—that is to say, System 1—than it might otherwise have been, because speakers were not obliged to give up the way they understood the world and its operations, its causal and spatial connectedness and the formal patterns that linked its various phenomena. In short, by converting to totally conventional symbolic signs, humans did not need to abandon the natural signs they had long recognized in their umwelt and had themselves so long used in the form of gestural and vocal indices and icons. For example, a two-sentence message such as "Come and see the marks in the sand by the river. They look like bear tracks." combines indexical function (in the imperative first sentence) and iconic function (in the declarative second sentence). Of course the person who uttered those two sentences might have been lying, or perhaps the hearers of these words might feel no pressing need to test their truth. Yet this accommodation of index and icon within symbolic discourse let hearers assure themselves that they could potentially test the truth of that statement by seeing for themselves whether what was asserted was actually "there" and whether what was attributed to it was correct.

Language play at the level of the sentence is itself a mimetic performance, one in which conventional signs (symbols) mimic natural signs (indices and icons). This imitative play is a charade in reverse: instead of substituting indexical and iconic signs for words, language substitutes words for indices and icons and derives rhetorical power from its capacity to step aside, as it were, and let the old prelinguistic sign functions take center stage. When it does so, it recruits visual schemas stored in semantic memory and through the simulation of visual perception—i.e., imagination—accomplishes the work of meaning-making.

Verbal Visuality: The Simulation of Perception

"Imagination" is not a term one often encounters in studies of cognitive linguistics, or in the sciences generally. It has lost the philosophic sense it had in the Romantic era when it was also associated with the claims of poets to reveries, waking dreams, and phantasmagoric visions. Nowadays, scientists and philosophers become scientists and philosophers precisely by reining in such mental excesses, and, if they have ever been subject to unbidden images, they are wise enough to keep that to themselves. Most people, moreover, when asked if they have vivid mental images, report that they remember having had them as children but that now they experience images only as the momentary sketchy forms that sometimes accompany thought. Yet mental images continue to have a function. Though the "pictures-in-the-head" version of imagination has proved untenable, verbal artists do provide us with quite detailed semblances, a fact that demonstrates our mind's capacity to organize complex visual representations in response to skillfully arranged suggestions. Language can indeed muster all the resources necessary to evoke in us images that are far more sustained and detailed than any of those we independently form. In his treatise on rhetoric, Aristotle recommended that the orator learn to "make things seem to happen before the eyes," *pro ommatôn poiein* (*Rhetoric*, 3.11.1). In this respect, as in others, the rhetorician's art is simply to exploit the properties inherent in language in order to enhance the persuasiveness of an argument.

Since vision is by far our most discriminative sense and our most relied upon means of locating dangers and benefits in our environment, it is not surprising that it plays such a major role in language. "Do you see what I see?" is just the sort of question that language seems designed to ask and to answer. Being concerned with asking questions about such questions, a growing number of linguists in the 1980s began to correlate visual-perceptual functions with the structures of language (Fauconnier, 1985; Jackendoff, 1987; Lakoff, 1987; Langacker, 1987). In their early work, these cognitive linguists drew from models provided by Gestalt psychology and by that branch of cognitive psychology that through empirical testing had determined that mental imaging replicates the basic procedures of visual perception. Leaders in this research included Robert Holt, Allan Paivio, Stephen Kosslyn, and Roger Shepard. (For a review of their research and its relevance to literary studies, see Collins, 1991a.)

Before I consider some of their insights into verbal visuality, I will suggest how several optical processes may have found their way into grammar.

Nouns signify those third-person entities upon which our knowledge of the world is founded. As verbal/visual symbols, they are modeled on the objects of focal perception, the figures that the eyes isolate from their peripheral grounds. In speech, therefore, a noun represents a fixation point that the verbally prompted mind momentarily dwells on before shifting to another object in imaginary space.[4] However, in adapting visual perception for its communicative code, language could not possibly reproduce all a scene's visually perceived features. As one in a series of simulated fixation points, a noun is, therefore, not quite equivalent to a visual fixation. Any perceived object important enough to be a figure isolated from its ground will probably be the target of multiple fixations. In normal speech we might say, "I saw a leopard." We might say that noun with raised pitch and volume or with a tremor in our voice. But if language were to replicate the actual visual event, it would have to reproduce it somewhat like this: "I saw the spotted pelt, the claws, the teeth, the brow, the leopard, the teeth, the tongue, the leopard, the teeth, the spotted pelt, the claws," and so forth. When perceiving an object, our eyes send to the visual center a rapid series of recurrent fixations. Later, when narrating this perceptual event, we name an object within a sequence of usually nonrecurrent nouns.

If nouns represent visual fixation points, *prepositions* may be used to represent *saccades*, those rapid shifts of optical focus from one fixation point to another. They also specify other eye movements that may accompany saccades—e.g., *vergent* movements, which set the focus at different points in depth, *conjugate* movements, in which both eyes turn together to examine objects arrayed at the same distance, and *pursuit* movements, in which the eyes follow a moving object in a smooth glide. Prepositions are not themselves visualized, because they represent saccades, which, due to the phenomenon of "saccadic suppression," go undetected by the viewer.[5]

Within a given context (e.g., a story about an encounter with a leopard), the noun "leopard" denotes a particular animal viewed from a particular perspective. In English it becomes "*the* leopard," not just "*a* leopard," which would mean it is only one of a genotypic class. In other informational contexts, common nouns may lose their deictic immediacy and need *modifiers*. If one says, "Give me the *big* stick," the adjective "big" is used to distinguish contrastively one stick from other sticks that are less big. In a present context, a demonstrative adjective may be used together with a deictic gesture, "Give me *that* stick," or a prepositional phrase may be used to modify the main noun, as in "Give me the stick *next to the tree*."

In an absent context, even when episodic details and adjectives abound, concrete nouns are not *concrete* in the sense of individuated and unique. A stick, in all its hefty, knobby, corticate quiddity, in no way resembles a noun. Speech as a means of evoking an absent object can do so only by referring to its class and prompting the addressee to visualize a prototypical representation of a member of that class. Obviously no language could have a different word for every individual object in its environment. It has to generalize, but generalization and category formation, as important as they are in language, are not creations of language. Like their primate ancestors, the first prelinguistic hunting-gathering humans had their minds well stocked with appropriate target images, field-guide-quality prototypes of plants and animals. The generalizing skill of imagination, preadaptive to lexicon and syntax, had to have preceded speech by many millions of years.[6]

In 1999 Ronald Langacker published *Grammar and Conceptualization*. In it he clarified some of his earlier analyses of language as, in part, a representation of visual experience. In his seventh chapter, "Viewing in Cognition and Grammar," he used the word "view" as evidence of the way we commonly speak of both our statements and our visual observations as having "point of view," "perspective," and "focus" and went on to outline a "variety of ways in which *perception* and *conception* can be regarded as analogous" (203, author's emphasis). For example, when we view an object of perception and it becomes our focus, we are also usually aware of other nearby ("onstage") objects but are less acutely aware of objects in the wider periphery, the "maximal field of view." Similarly, when we conceptualize a notion, either by speaking, hearing, or thinking it, this notion becomes our "profile," i.e., the designatum that stands out from its base, the latter comprising immediately associated concepts and those more distantly related (204–5).[7]

Langacker then proceeds to consider concepts ranged along a spectrum of specificity, such as "thing→object→vehicle→car→Dodge→Dodge Colt." This is analogous to the spectrum of "enhanced visual acuity we

TABLE 6.1 Figure–ground differentiations as reflected in language

FIGURE	GROUND
An entity of undetermined spatial (or temporal) properties, but one that is assumed to be:	As a reference entity, having known properties that can characterize the figure's unknown properties, it is:
• More mobile	More permanently located
• Smaller	Larger
• Geometrically simpler	Geometrically more complex
• More recently on the scene or in awareness	More familiar or expected
• Of greater concern or relevance	Of lesser concern or relevance
• Less immediately perceivable	More immediately perceivable
• More salient, once perceived	More backgrounded, once figure is perceived
• More dependent	More independent

experience in approaching a distant object: the closer we get, the better we see it (the smaller features we are able to detect and resolve). Viewing distance correlates not only with acuity (a negative correlation) but also with scope (a positive one)" (206). That is, in visual perception and in verbally prompted imaging, a scene may be composed both of clearer, closer objects and of objects that, being more distant or peripheral, are less clear, a placement of objects that generates the figure–ground relationship.

In table 6.1, Leonard Talmy (2000:315–16) carefully differentiates figure and ground as this principle appears in visual and verbal experience. These contrasting characteristics would be perfectly consistent with the perceptions of a hunter-gatherer, alert to pinpoint the location of a rabbit or an edible root plant. It is also the way we normally describe objects in space. Talmy asks us to consider this example: *The bike is near the house.* "Near" is a reciprocal relationship, so if the bike is near the house, we should logically be able to say, *The house is near the bike.* But that last sentence is anomalous. The bike, being movable and smaller, is the best candidate for figure status and should therefore be the subject of the sentence; the house, being a larger, more permanent object should be its ground and function within the predicate. The terms *target* and *landmark* are virtual synonyms of figure and ground, but because they imply serial activity (search and wayfinding), they have seemed appropriate in describing the serial processing of language. Accordingly, the bike is the target, while the house is the landmark. It is not likely that anyone would use the bike as a landmark by which to locate the house.

On the other hand, we can, if we wish, defamiliarize a visual description by reversing the normal figure–ground, target–landmark relationship. William Carlos Williams did just that when he placed a "red wheelbarrow / glazed with rain / water," a movable but here a static, larger object, "beside the white chickens," an ever-moving set of smaller objects. Functioning as target and landmark, his phrasing is equally anomalous: the chickens do not help us perceive the position of the wheelbarrow, unless he is referring to a painting, an art form that can also play tricks on our perceptual habits.

In visual perception, figure is determined by central vision and ground by peripheral vision. Together, figure and ground constitute a dyad, each complementing the other. But considered separately, each represents images at different degrees of resolution. If we apply this difference to mental imaging generally, then to verbally cued images, we have further support for Langacker's claim that the specificity of nouns correlates with their imagined distance from the viewer. For example, if verbally prompted to visualize a maple leaf, I would report a clearer image than I would if prompted to visualize a maple *tree*, which would be a much more acutely resolved image for me than a maple *forest*. If I were actually perceiving these three items, at each step in the process my visual scope would have to expand. The leaf at arm's length might fill a circular field of 25 degrees and be comfortably viewed in central vision. The tree, viewed at a distance of 50 feet, might occupy 100 degrees of optical field, thereby overflowing into my peripheral field and requiring multiple saccadic shifts to scan its base, trunk, boughs, and uttermost leafy extensions. Needless to say, my visual perception of a maple forest would likely necessitate a scope so broad as to overflow even my peripheral field.

The link between visuality and language involves, as Langacker put it, an analogy of "*per*ception and *con*ception," a correlation that Leonard Talmy (2000) boldly contracted by deleting the italicized prefixes and proposing the "idea of a single continuous cognitive domain, that of 'ception'" (102). Here, Talmy implies that, like violet and red in the color spectrum, perception and conception are distinguishable but not radically separable. In the light of his proposal, it is useful to consider how language behaves on its continuum, how, as a window to the mind, it reveals the mind in the act of "ception." Archibald MacLeish's famous little poem, "Ars Poetica," demonstrates this in several passages, but I will cite just four lines (17–20):

A poem should be equal to:	17
Not true.	18
For all the history of grief	19
An empty doorway and a maple leaf.	20

If we take the first two lines as a proposition and the next two lines as its illustration, we understand these two couplets as linked to one another by a kind of equal sign, a relation of equality that is, paradoxically, "not true." When we examine the illustrative second couplet, we find it composed of yet another implied equation: (1) "all the history of grief," a visually deficient, but affectively powerful conception = (2) "an empty doorway and a maple leaf," a pair of visually acute but affectively neutral simulations of perception. Line 19 presents a concept, "the history of grief," that could only exist in the medium of language, the symbolic system that, as Deacon pointed out, allows us to time-travel, to make promises, and, in this case, to feel loss and regret. Line 20, however, represents images that are not language dependent and could have been signified indexically or iconically—e.g., by pointing to a doorway and a leaf, drawing pictures of them, or sequencing them as stills in a film montage. The distance on Talmy's continuum between abstract and concrete "cepts" can be great, but sometimes, as in this case, it can prove astonishingly bridgeable.

As George Lakoff, Mark Johnson, and Mark Turner have demonstrated over the past quarter century, metaphor is a pervasive cognitive phenomenon based on the largely unconscious association of separate conceptual domains that find expression in thought, action, and language. When, on the conscious level, it is linguistically expressed as a sometimes novel equation of words from distant semantic domains, we recognize it as a trope (Lakoff and Johnson, 1980; Lakoff and Turner, 1989). In its typical verbal form, a word or phrase (A) that represents a concept that is less familiar, less readily comprehended, is equated with a word or phrase (B) that represents a percept that is more familiar, more readily comprehended. Even if this target (A) is a concrete object, metaphorizing it indicates that it is the speaker's *understanding* of that object that needs to be clarified. For example, in the well-worn metaphor "My surgeon is a butcher," *my* surgeon, the target, is both the subject of the sentence and the focus of attention. From the addressee's point of view this specific surgeon is probably a less familiar item, probably a total unknown, whereas the source term (B), *a* butcher, is much more readily comprehended. Similarly, when Burns said "My love is like a red, red rose" he was talking about a specific concrete woman whom we don't know and perhaps not even he finds wholly comprehensible. The red rose—that's a lot easier for all parties to understand.

The two elements that come together in a metaphor are processed in parallel, a fact that suggests that metaphor is a linguistic dyad. But language, being a serial medium, keeps the verbalized components separate, a condition that prompts us to differentiate target and source by assigning broad awareness to one and narrow attention to the other. This would be a mistake, however, for when they blend into a metaphor, target and source, while still distinguishable, are no longer separate entities. When the metaphor dyad happens, the target is transformed: its many concomitant, unselected details become a ground that surrounds a figure of focally selected details. Thus, when Romeo declares that "Juliet is the sun," the concomitant facts that she is also a willful thirteen-year-old girl, lives in fifteenth-century Verona, and belongs to a prominent family that is the sworn enemy of Romeo's kin—all this context remains, but it is momentarily peripheral to her "solar" details, her warmth, brightness, and centrality in Romeo's mind. This and every other verbal metaphor forms a dyad, which, I submit, is the same verbal pattern that Ezra Pound (1913/1974:37) called an "Image," describing it as capable of presenting "an intellectual and emotional complex in an instant of time. . . . It is the presentation of such a 'complex' instantaneously which gives that sense of sudden liberation; that sense of freedom from time limits and space limits; that sense of sudden growth, which we experience in the presence of the greatest works of art."

As I just mentioned, the most important component of a metaphor is (A), the intended target of this tropological maneuver, whereas (B) represents a semantic domain, a source, from which certain applicable properties are transferred to (A). Of course, the potential of a metaphor depends on the aptness of the source domain but this potential is actuated only when those two disparate concepts become momentarily fused. Both "butcher" and "rose," while representing perceivable objects, do not, however, represent *specific* objects. As metaphor sources they are generic types—conceptual categories.

As these examples illustrate, metaphor reproduces in the medium of language the visually perceived relation between a *target* and its *landmark*, a relation governed by the figure–ground principle. When, however, this organizing principle is transposed from visual perception to visual *conception* as we enter the realm of verbal visuality, there is a significant difference. At the perceptual pole of Talmy's "ception" continuum, target+landmark and figure+ground each establish a common spatial context. They belong together. But at the conceptual pole the two verbal components of a metaphor, belonging as they do to semantic domains quite distinct from one another, have no such common context. They do, however, have one thing in common, the speech event in which they momentarily come together. Langacker (2002) implies as much when he says that every "speech event, its participants, and its immediate circumstances" constitute its *ground* as viewed from the speaker's perspective. Just as a visual ground extends outward from the focalized figure at two degrees of separation—first, the "onstage" objects associated with the figure and, second, the peripheral objects in the "maximal field of view"—so too the speech event indicates the proximal ground through deictic expressions, such as *I, you, here*, and *now*, and the more distant ground by expressions such as *yesterday, tomorrow*, and *last year* (2002:318–19). To understand how metaphor is grounded, there may be some value in stepping back from it a bit and considering the entire two-part trope as a figure and its surrounding context—the speech situation, the sentence, and the complete utterance or text—as its ground. In so doing we recognize the speech situation as the standard model for the performance of verbal art in both its oral and its written settings.

Verbal Visuality: The Simulation of Action

Perception, especially *visual* perception, has a certain calm about it, especially when the viewer, securely situated, observes objects in their spatial and temporal relations to one another. From that vantage point, one can leisurely parse the scene, distinguishing figures from grounds and applying one's prior knowledge to each selected component. Action, however, entails the viewer's physically entering and engaging in that three-dimensional world of objects and, with that, a shift from the allocentric to the egocentric frame of reference.

Informed by their readings in Gestalt and cognitive psychologies, early cognitive linguists absorbed and reformulated a wealth of crossdisciplinary data and models of visual mapping. They were slow, though, to incorporate contemporaneous discoveries concerning the structures of the visual brain. Somewhat outside the cognitivist mainstream, but well respected for his work on syntax, Talmy (Thomas) Givón was an exception. In *Bio-Linguistics* (2002), he applied the dual-stream model of the visual brain to the structure of language:

> The two streams of visual information processing correspond, rather transparently, to two core components of human cognitive and linguistic representation systems . . . , of which they are but the visual precursors:
>
> > object recognition = semantic memory = lexicon
> > spatial relation/motion = episodic memory = propositions.
>
> This identification is surprisingly straightforward. First, recognizing a visual object as member of a generic type according to its visual attributes is but an earlier prototype of lexical-semantic identification. Second, processing an episodic token of spatial relation between objects is but the early prototype of episodic-propositional information about *states*. And processing an episodic token of spatial motion of one object relative to another is but the early prototype of episodic-propositional information about events. (Givón, 2002:136)

This quote warrants closer examination. The two visual streams (or pathways), the ventral and the dorsal (see pp. 90–93), operate together to provide the brain with two different kinds of information—namely, object recognition and spatial location. As such, they are the evolutionary precursors of the "human cognitive systems," semantic and episodic memory. Following Endel Tulving's (1983) analysis, Givón understands these two memory systems as two distinct ways of organizing experience: (1) the semantic system, a memory archive of generic types, specially marked mental compartments into which newly perceived tokens may be placed and thereby identified; and (2) the episodic system, an archive of autobiographical events, a series of episodes having happened in particular places and particular times and tagged with particular emotional overtones. The ventral stream, via the semantic system, has generated the lexicon, while the dorsal stream, via the episodic system, has generated propositions and the syntax that governs them.

In arriving at these conclusions, Givón was relying on the early work of Ungerleider and Mishkin (1982). If he had taken into account the work of Milner and Goodale (1995; see chap. 4), he might have made an even more impressive correlation of visual neuroscience and linguistics. The latter researchers' main contribution had been their discovery that the dorsal stream, initially termed the "where" path, was better characterized as the "how" stream. He would have learned from them that this dorsal stream processes the ever-changing optic flow, the visual part-perceptions that objects present as the viewer moves among them. This neural stream processes perceptions, but perceptions *for online action*, and in doing so projects them into an egocentric frame of reference. That is, the dorsal stream is not tasked with object recognition, or, as Givón put it, the "spatial relation between objects," but only with object location in *relation to the moving viewer*.

When Givón says that what the dorsal stream processes is "an episodic token of spatial relation between objects [that] is but the early prototype of episodic-propositional information about *states* [i.e., spatial relations]" (2002:145), he seems to be referring to that *later* stage at which an experience, now stored in episodic memory, is retrieved and reconstructed in a propositional format that a hearer or reader can more readily understand. When, for example, we tell someone about a personal episode, we do not try to convey the events exactly as we experienced them, especially if they involved our rapid reactions to emergent circumstances (cf. the encounter with the leopard mentioned earlier). We need to set the scene, to reassemble the elements in it, the "states," as the ventral stream might have organized them, if it had had the time to project a calm, allocentric view of those elements.

That the brain can simulate action, as well as perception, has long been assumed, but not until recent decades has this assumption been empirically established. A significant support for this idea came from the discovery in the 1990s of mirror neurons in the brains of primate monkeys. Since then, brain imaging has presented strong evidence for their presence in humans as well (see pp. 46–47). While psychologists, of course, were the first to recognize the potential importance of such findings, cognitive linguists were soon after attracted to the possibility that the Mirror Neuron System could help explain how language represents the visual imagery of action.

In a notable collaboration, the mirror neuron researcher Vittorio Gallese and the cognitive linguist George Lakoff teamed up to write a paper entitled "The Brain's Concepts: The Role of the Sensory-Motor System in Conceptual Knowledge" (2005). For Lakoff, concepts had been a central concern since the 1980s when he and his colleagues began examining metaphor as a means by which concepts emerge from everyday experiences. When the Mirror Neuron Hypothesis was announced, he saw in it converging evidence to support that central premise of cognitive linguistics, that concepts *and* language, rather than being products of symbolic logic or of an autonomous system of signs, originated in perception and action and are therefore fully embodied entities.

In their paper, Gallese and Lakoff use the word "simulation" to characterize both covert perception and covert action. It therefore corresponds to what one might term "mental mimesis" or, simply "imagination." When they declare, "*Imagination is mental simulation,*" they add that this is "carried out by the same functional clusters used in acting and perceiving" (458, authors' emphasis). So: imagination = simulation = visuomotor cognition. Language conveys "*all* concrete concepts—concepts we can see, touch, manipulate." But are these the only concepts it conveys? What about abstract concepts? For this they refer to the theory of conceptual metaphor: "Each conceptual metaphor is a mapping across conceptual domains, from a (typically) sensory-motor source domain to a (typically) non-sensory-motor target" (469–70).

But this only provokes a follow-up question: How did the non-sensory-motor domain come into existence in the first place? Oddly enough, the authors neglect to tell us. They could have easily done so, however. They could have invoked the Lakovian theory that abstract concepts such as love, death, strife, etc., each came into being through an association of embodied experiences, perceptual and motoric, experiences that in the form of image schemas continue to support each concept by their metaphorical input.[8] The authors do introduce that old standard LOVE IS A JOURNEY, with all its references to "bumpy roads," marriages "at the crossroads" or "on the rocks," etc., but what they do not say is that it is language that supplies words for those "functional clusters used in acting and perceiving" (458) and that language, though it is an embodied adaptation of prelinguistic cognition, is nonetheless a system of symbolic, therefore *dis*embodied, signs.

Though I have chosen to examine simulations of perception first, then simulations of action, it would be a mistake to dissociate the former from the latter: the ventral stream that subserves perception communicates with the dorsal stream that directs action (Goodale, 2011). They are parallel functions and, though separate, are like a securely married couple that pool their income and keep no secrets from one another. As Gallese and Lakoff's paper demonstrates, it is impossible to address the simulation of action without discussing the simulation of visual perception. If we ask how language prompts us as hearers (and readers) to simulate action, and apply the Mirror Neuron Hypothesis to this question, we may infer that when we imagine another human performing a particular act, we also simulate that action. That is, language prompts us, first, to retrieve image schemas that represent a person moving in space and then, by retrieving our own action schemas, we understand the action by covertly replicating that person's movement. We first imagine we see in an allocentric frame a human in motion and then imagine that we *are* that human and that we see the world egocentrically framed from that viewer's perspective.

Supplementary to the Mirror Neuron Hypothesis, the Dual-Path Theory of visual perception suggests further implications. Simulated visual perception through language would evoke ventral stream processing and produce the recognition and selection of figures from grounds. Here we note an interesting fact: language reverses the order of visual parsing. The recognition process now precedes selection, for as soon as a noun is said (or read) its semantic content is supplied (it is recognized as X or Y) and now all that is left is to imagine its perceptible features (to select it as a bounded figure from its ambient ground). The motion of objects, human or otherwise, would also be first processed through a simulation of the ventral stream (not the dorsal) and linguistically conveyed by verbs and prepositions. As Milner and Goodale (1995) have maintained, the visual information that the dorsal stream provides the moving observer is largely peripheral and preattentive. Therefore, when we respond to a sentence by feeling ourselves inside that third-person body and simulating that person's movements, that dorsally simulated experience is real, though difficult to specify.

So far, I have been examining the capacity of language to represent the actions of objects in a simulated visual field. A description of such an event is assumed to be an intentional sharing of real-world information. There are, however, other forms of expression that seem less deliberately selected but nonetheless inherent to language. One of these is "virtual," "subjective," or, as Talmy (2000) termed it, "fictive" motion. As Talmy has strongly argued, the same prepositions used to link targets and landmarks can, when modifying certain action verbs, represent nonmoving objects as though in motion. As an example, he offers this sentence: "The mountain range goes from Canada to Mexico" (2000:104). He also speaks of "orientation paths," e.g., prospect paths ("The cliff wall faces toward the sea") and targeting paths ("I pointed the camera into the living room").

In his discussion of fictive motion, Talmy later proposes another correlation between vision and language. Both combine two distinct subsystems: *content* and *structure*. Visual content is registered as precise, absolute, Euclidian percepts. It therefore corresponds to the allocentrically framed visual perception associated with the ventral stream. Visual structure, on the other hand, is registered as the rough-and-ready assessments we make of the relative positions of objects in topological space as we physically navigate in the object world. "Structure" therefore corresponds to the egocentrically framed optic flow associated with the dorsal stream. "Content" is linguistically conveyed by open-class forms (lexical words, i.e., nouns, verbs, adjectives, and adverbs), while "structure" is conveyed through closed-class forms (functions words, i.e., pronouns, prepositions, conjunctions, and articles) (Talmy, 2000:160–68). It is the structural subsystem of vision that is implicated, he suggests, in the linguistic phenomenon of fictive motion.[9]

The experience we have of fictive motion in language, Talmy says, may be explained in several ways. First of all, it may be regarded as either observer neutral or observer based. If the latter, it is the person whom the reader imagines as observing the motion that either moves or visually scans an object's apparent trajectory. If, however, we interpret an object's motion as its "emanation," Talmy says we must take an observer neutral approach. But why identify the motion of this referent as an "emanation?" His only explanation is that "this category [emanation] appears to have been largely unrecognized" (105) and later adds that it may be connected with the worldwide belief in ghosts and their power to pass through walls and other obstructions (126).

Showcased in a collection of previously published papers by an eminent linguist, this intriguing chapter was initially well received (Turner, 2002; Coulson and Oakley, 2005; Wallentin et al., 2005).[10] Not surprisingly, however, talk of disembodied, unmoved movers haunting the corridors of language has aroused the attention of cognitive linguists for whom embodiment is the watchword and whose first principle is "that any aspect of space that can be expressed in language must also be present in nonlinguistic spatial representations" (Landau and Jackendoff, 1993:217). For them there were better and simpler explanations for fictive motion, including mental scanning as indicated by measurable eye movements and motor simulation as indicated by brain imaging (Matlock, 2004; Richardson and Matlock, 2007).

In a paper that appeared in 2009, Line Brandt went further and questioned the broadness of the fictive/factive (or figurative/literal) contrast that Talmy makes, proposing instead that all sentence construals are mind dependent and inevitably share a subjective character. She quotes an example: "The fence goes from the plateau to the valley" (Brandt, 2009:580). Of this Talmy had written, "[T]he literal meaning of the sentence fictively presents the fence as moving" (Talmy, 2000:101). To this she replies that the "conceptualizer does not see the fence as moving; rather, the motion is played out in the mind of the conceptualizer in constructing the representation of the (stationary) fence" (Brandt, 2009:583). Fictive motion is simply a function of the conceptualizer's process of representation, an act of simulation that can never be "observer neutral."

To support her view, Brandt cites Langacker, who, like Talmy, has for several decades also been studying this interesting phenomenon. In an article published in 2001, he stated that "[c]onceptualization is inherently *dynamic*. It resides in mental processing, so every conception requires some span of *processing time*—however brief—for its occurrence" (Langacker, 2001:8 author's emphasis). Language is a serial medium and, though a given "temporal order of words encourages us to activate conceptual elements in a particular sequence, . . . [w]e have every reason to suspect that processing proceeds simultaneously on multiple time scales, and with respect to numerous parameters" (11). It is the hearer (or reader) who processes the words, their temporal meanings within this temporal medium. The possibility, therefore, that a multiple of different time scales may be processed in parallel suggests a subtlety that enhances, as it exceeds, the complexity of fictive motion.

Whatever reason Langacker had back in 2001 to "suspect that processing proceeds simultaneously on multiple time scales," evidence has emerged that multiple timing is indeed a factor in the brain's massively parallel networks. In 2012 an international team of researchers led by Joerg Hipp published the results of a study, based on magnetoencephalography (mercifully shortened to MEG), that concluded that every network maintains its cellular connectedness by synchronizing its firings at a particular frequency. For example, cells in the hippocampus associated with episodic memory are active at around 5 hertz. Perception and action networks fire at frequencies of from 32 to 45 hertz, while other networks were found to range from 8 to 32 hertz (Hipp et al., 2012). This means that, though the networks may be widely distributed in space and resemble those untamably entangled cables behind our computers, they are actually marvels of temporal order, like the stations on a radio dial, each with its own assigned frequency. Not only does this finding appear to support Langacker's intuition, but it may also support Merlin Donald's (2007b) hypothesis of intermediate-term episodic consciousness and prove relevant to the study of pitch-based metrics in music and verse.

The dynamic character of brain networks strongly suggests that any linguistic theory of fictive motion should begin, and plausibly end, in biophysical analysis. This axiom also applies to the motoric structure implicit in the sentence—the progression from subject to predicate, or topic to comment. As a way to understand the prelinguistic origins of this bipartite structure, the German linguist Manfred Krifka (2007) borrows Yves Guiard's (1987) analysis of bimanual coordination (see pp. 39–40). Any utterance is assumed to be about something—namely, its topic.

> [T]he aboutness topic "picks up" or identifies an entity that is typically present in the common ground of speaker or hearer, or whose existence is uncontroversially assumed. This corresponds to the preparatory, postural contribution of the non-dominant hand when it reaches out and "picks up" an object for later manipulation. The comment then adds information about the topic, which in turn corresponds to the manipulative action of the dominant hand. (Krifka, 2007:83)

Krifka then elaborates this parallel with a number of striking observations concerning the dyadic asymmetry of both manual coordination and sentence structure:

> [T]he actions of the non-dominant hand typically precede the corresponding motions of the dominant hand in bimanual manipulations. This directly corresponds to the typical temporal order in which topic/comment-structures are serialized, with the topic being mentioned first, and then elaborated by the comment. A second point of similarity concerns the scale of motion. We have seen that the motions of the non-dominant hand are more coarse-grained, whereas the motions of the dominant hand tend to be on a more fine-grained scale, both spatially and temporally. In addition, the movements of the dominant hand are more frequent, and generally expend more energy. This is related to the realization of topic/comment structure, where the topic tends to be de-accented, and the comment typically bears more pronounced accents. Furthermore, notice that the prehension of an object by the non-dominant hand is, in a sense, static, as it does not affect the internal nature of the object. This is only done by the manipulation of the object by the dominant hand. Quite similarly, identifying a topic does not change the information state yet, but only prepares a change; the change itself is executed by the comment. (2007:84)

Language, as we have come to know it, is a medium of visual representation with the capacity to prompt addressees to simulate (1) the perception of identifiable objects based upon image schemas and (2) the sense of acting upon those objects based upon motor schemas. As speech, itself a motor behavior, language is also aligned with the routines of eye–hand coordination (from which it may have derived its topic/comment structure) and with the step-by-step seriality of tool making and tool use.

As I have tried to show in this chapter, language incorporates in its structures a number of prelinguistic visuomotor routines that it uses to represent our human umwelt and share our knowledge of it. This, we might say, serves the purpose of disseminating object knowledge. But then there is that older, underlying motive: the need to possess and advantageously use social knowledge. I think it is fair to say that object knowledge is most widely valued not so much for itself as for the social advantages it confers upon its possessors. I will therefore close this chapter with comments on the "rhetorical imagination," i.e., the means by which rhetoric communicates object knowledge through a persuasive use of visual imagery.

The Rhetorical Imagination

Classical rhetoricians recognized five steps, or canons, in the composition of an oration. Other divisions in other oratorical traditions might be equally useful, but I will stick to this tradition for purposes of convenience. The five are (1) **invention** (L. *inventio*, Gr. *heurêsis*), the finding of useful topics; (2) **arrangement** (L. *dispositio*; Gr. *taxis*), the order in which the selected topics are set forth; (3) **style** (L. *elocutio*; Gr. *lexis*), the level of language and figurative devices used; (4) **memory** (L. *memoria*; Gr. *mnêmê*), the ability to speak without notes and thereby appear strongly motivated by conviction; and (5) **delivery** (L. *actio*, Gr. *hupokrisis*), skillful use of acting technique, including appropriate vocal and gestural features.

Rather than simply setting forth these canons of rhetoric as an artful speaker's skills, I will treat them as forms of social negotiation through which a first-person addresses some second-person(s) in order to formulate a consensus attitude toward some third-person(s)—other humans, animals, things, situations, etc. Accordingly, I will assume that each canon entails a cognitive process that to some extent both speaker and audience experience during the speech. I will also assume that rhetorical techniques are also used in spontaneous speech events, not only in formal preplanned speeches, and that the rhetorically adept were always able to "think on their feet" and improvise persuasive arguments. Finally, in discussing each canon, I will suggest that its power derives from a past still deeply rooted within us in those portions of the brain our prelinguistic ancestors depended upon to communicate with one another.

Invention is the skill of finding the most persuasive things to say in order to make one's case. As with any search, one has a better chance of finding one's targets if one knows where to look. Hence traditional rhetoric refers to places (L. *loci*, Gr. *topoi*) where useful arguments are to be found. I will refer to the contents of *topoi* as "topics," because it is a broader term applicable to speech, including speech that is not explicitly argumentative. "Topic" is also an appropriate term because a *topos* functions very much like a topic in a sentence: it introduces a commonly accepted idea to which the speaker adds a special comment—in this case, an elaboration that brings into focus a situation of current concern. These locations were referred to as "common places" because the topics found there could be used to support a wide variety of arguments.[11]

When the character called "Chorus" appears at the beginning of Shakespeare's *Henry V*, he starts with a rhetor's prayer of sorts: "O for a Muse of fire, that would ascend / The brightest heaven of invention, / A kingdom for a stage, princes to act / And monarchs to behold the swelling scene!" By the mid-eighteenth century any "Muse of fire" would be associated with "imagination," but for Renaissance poets she was the inspirer of "invention." He goes on to name his theme: "Then should the warlike Harry, like himself, / Assume the port of Mars." If time travel were possible and if the staging could be to scale, then a historical exemplum (common place #1) would be King Henry, who would appear as he was *and* as a mythological exemplum (common place #2), the god of war. This prologue then proceeds to other common places, mainly based in contrasts of scale (e.g., this theater as an "unworthy scaffold," etc.).

Finding usable things in familiar places suggests a broader analogy: language as a searchable landscape. The topics it is stocked with can be put to various uses and are therefore like the stuff of improvised tools, such as stones, branches, and other natural objects. If verbal artifacts are *made* tools, then rhetorical topics are *found* tools and the rhetor a resourceful *bricoleur*. Spontaneous speech production, even when unskillfully performed, does indeed resemble a searching for words that are there somewhere and just need to be located, inwardly found words that, once found, extend the outwardly persuasive influence of their user.

The second canon, **arrangement**, is the order in which the speaker visits the common places and retrieves topics. In a formal speech, this sequence is tactically chosen, but, even in a spontaneous speech, attention is paid to the order of details. In either case, the hearers follow the speaker as they would follow a guide through a dense forest, pleased, sometimes marveling, at his or her skill in pathfinding and bushcraft. This action of following a leader through a landscape is implied by the Greek term *diêgêsis*. This was storytelling, as distinct from *mimêsis*, a story*showing* through enactment. Rhetorical diegesis, for an audience, meant following, at times anticipating, the leader's discovery of topics. Though commonly defined as a narration of events, diegesis also necessarily included the description of persons and settings.

Following this searcher's route is possible, however, only when the attention of the audience is riveted on the process. This is the function of **style**. As the speaker's means of maintaining that attention, style also involves avoiding all irrelevant shifts of attention occasioned by misused or dysfluent language, diction inappropriate to the subject matter, or any other verbal distraction from the performance. On the positive side, style lends speech the power to evoke mental imagery in the audience, what Aristotle spoke of as placing things "before the eyes." The Chorus in *Henry V* urges his audience to let the actors work on their "imaginary forces," the *vires imaginativae* of medieval "faculty psychology," to augment their outward visual perceptions. Ordinarily actors have an advantage in this matter since their physical movements hold an audience's attention far better than the mere words of any orator or storyteller, who, as a diegetic performer, must evoke the visual absent as though visually present. As turns or twistings of language, tropes are the orator's and writer's best means of prompting others to visualize the unseen. Seeing may be believing, but imagining can also be an effective persuader.

These rhetorical twists often play with irrational forms of visual experience. One such trope is *apostrophe*, a turning-away-from one's audience in order to address a person, thing, place, or idea, as though it were visible to the speaker—a hallucinated image calculated to demonstrate the extreme emotion of the speaker. Another is *personification* (Gr. *prosopopoeia*), literally mask making. This usually entails describing a nonhuman entity as expressing the human traits of consciousness and will. Melville's Captain Ahab exhibited the twistedness of personification in chapter 36 of *Moby Dick*, "The Quarter Deck." This passage reveals the basis of this trope as "theory of mind" (ToM), projected anthropomorphically onto nonhuman nature, a primitive turn of mind that his First Mate names "madness!":

> "Vengeance on a dumb brute!" cried Starbuck, "that simply smote thee from blindest instinct! Madness! To be enraged with a dumb thing, Captain Ahab, seems blasphemous."
>
> "Hark ye yet again,—the little lower layer. All visible objects, man, are but as pasteboard masks. But in each event—in the living act, the undoubted deed—there, some unknown but still reasoning thing puts forth the mouldings of its features from behind the unreasoning mask. If man will strike, strike through the mask!"

Here, of course, Moby Dick is not simply wearing a mask, a *prosôpon*. Instead, in all his cetaceous bulk, he embodies the mask behind which some other malevolent agent lurks. As a way to grab the attention of an audience, a prosopopoeic descent to this "little lower layer" where things may be observing one and consciously plotting can, at the very least, produce an uncanny effect.

In the theory of style, however, the most important tropes remain *metonymy* and *metaphor*. Metonymy, literally cross-naming, substitutes one noun for another noun with which it is physically associated—e.g., a tool for a tool-user ("pen" for writer and "sword" for soldier), container for contents (to drink a "glass"), a building for its occupant (the "White House" for the U.S. president), or a part for the whole ("foot" or "boots-on-the-ground" for infantry). As these standard examples indicate, metonyms, like metaphors, have a way of becoming the idioms of ordinary language. Though language is the graveyard of dead metaphors and metonyms—fossil poetry, as Emerson once called it—living tropes can still sprout from its soil.

According to Roman Jakobson, metaphor and metonymy are based on two different yet complementary ways of connecting thoughts: by similarity and by contiguity. In this, they correspond to those two presymbolic sign types, icon and index. Thus metaphor associates two symbolic signs by treating one (the source) as an iconic signifier of the other (the target), because it shares certain properties with its signified. Metonymy, on the other hand, associates two symbolic signs by treating one (the metonym) as the indexical sign of the other, as, for example, the cause or the effect, the part or the whole, of its signified (Jakobson and Halle, 1956; Silverman, 1983).

Long considered mere ornaments of speech, these two tropes are profoundly important to the structuration of language. At its simplest level, language is composed of different, meaningless phonemes that, when linked in certain sequences, form meaningful words. Thus the sounds [k], [t], and [a] may be rearranged to form "act," "tack," and "cat" (Hockett, 1960/1982). At a higher level of complexity, we can speak of whole words inhabiting a semantic universe at various degrees of relatedness to one another, each available to serve the speaker's needs by presenting itself for selection in a given sentence. These words exist in parallel within a quasi-three-dimensional world, the lexicon, but, if they are to be uttered (or written), they must enter into a two-dimensional serial state, discourse, where they must now obey the rules of syntax. As lexical entries (*lemmas*), words exist in what is termed a paradigm, a state of interrelatedness and difference, from which they are selected by a speaker and inserted into a linear series termed a syntagm. We might picture this word-filled paradigm as hovering above a horizontal line, the prospective utterance. Descending from the paradigm is a vertical line, the "axis of selection" down which one word at a time will be dropped onto the horizontal line, the syntagmatic series of slots, the "axis of combination" (Jakobson, 1960). Jakobson went on to point out that metaphor results when the "principle of equivalence" is projected from the axis of selection onto the axis of combination, i.e., when different but related words are treated *as though equivalent* and, as such, inserted within the linear sentence. Metonymy, on the other hand, results when the axis of combination dominates and connection is wholly determined by the way real-world signifieds relate to one another in space and time.

Metonymy and metaphor have their prelinguistic roots deeply embedded in visuomotor communicative codes. Metonymy, as we have seen, derives from indexical signs (e.g., gestures used deictically to direct the attention and movement of others), while metaphor derives from iconic signs (e.g., handshapes and pantomime). They are also grounded in the two visual processing modes associated with the ventral and dorsal streams. As a visual trope, metonymy is somewhat more problematic than metaphor. In its standard form it is always a two-part operation that mentions only the signifier, never the implied signified. For example, the "pen" stands for the unspoken "writer," the "sword" for the unspoken "soldier." Yet the two metonymic components exist together in the mind and their visual relationship is imagined as a searcher might process a landmark and its as-yet-unseen target. These conventional emblems might be called "dead metonyms." When located on Jakobson's axis of combination, however, metonymy becomes a much more interesting and vital stylistic device that now works by serial shifts of focus. An object is introduced, e.g., a *roof*, a word that immediately brings to mind a *house*, which in turn implies *human inhabitants*, and so forth. This is a process that Paul Deane (1992) identified as "spreading activation," a concept first formulated by Allan Collins and colleagues (Collins and Quillian, 1969; Collins and Loftus, 1975). If a concept is sufficiently familiar, its mention will prime a number of associative chains that the speaker (or writer) may use to lead—or lead astray—the hearer (or reader).

These metonymic shifts of focus, interpreted as simulated saccades, might help explain the phenomenon of "fictive motion." If we combine this saccadic simulation with simulated optical flow, we have a full simulation of dorsal stream processing. In other words, a metonymic series can represent a field of objects as they would appear to a viewer moving among them, progressing from landmarks to targets for the purpose of either mapping the space or searching for particular items within it. This might also help explain the relation of metonymy, the simple, standard form, to that strictly part–whole trope, *synecdoche*. If we regard the difference between these two as between two different modes of visualizing contiguity, we may then hypothesize that the difference between the standard two-part metonymy and its progressive, synecdochal variant is the difference between the verbally cued simulations of the ventral and dorsal visual streams, respectively. In what I have called the simple, or standard, metonymy (e.g., the pen standing for a writer, or the writing profession), the signifier relates to its signified allocentrically: here is A, now think of the often contiguous B. In a narrative, however, synecdoche, or synecdochal metonymy, often relates one object to another and that other to yet another and another, as some conscious focalizer progressively moves through a detail-rich, egocentrically framed space.

In preliterate cultures, the fourth rhetorical canon, **memory**, would be absolutely necessary for a speech that needed to marshal extensive evidence and counter anticipated objections. In a literate culture or one in which orators would be expected to be literate, one might suppose that speeches could be recited from scripts or notes. The reason literate orators normally committed a speech to memory seems to have been because spontaneous, albeit artfully spontaneous, speech is more persuasive. The sight of a speaker actually, or seemingly, thinking while addressing an audience, enacting the search for ideas, the invention of topics, stimulates members of the audience to participate in this quest. (Remember, the fifth canon, delivery, was termed *actio* in Latin and *hupokrisis* in Greek, both of which meant an actor's stagecraft.)

The mnemonic method taught in Roman schools of rhetoric has been called the "mental walk," or the method of *loci*. An orator would choose in advance a setting with which he was familiar, often a public building, such as a temple, and mentally map that interior with all its architectural details. Once that mental map was committed to memory, the orator could represent each topic by an iconic emblem, then project each emblem onto a particular place, a *locus*, on his mental map. Once these two sets of visual cues, the *loci* and the emblems, were firmly connected, he would be ready to deliver his outward talk while visualizing his inward walk (Yates, 1966). This method worked so well because it was a simulation of visuomotor perception that combined the actions of both visual streams. In walking toward each *locus*, the visualizer was simulating the action of the dorsal stream. As each emblem revealed itself, the simulation of the ventral stream took over, this object was selected, and then it was recognized as representing the topic that now needed to be presented. The Roman practice of the mental walk resembles the process of discovering concepts stored in "places," i.e., invention. But, though *loci*, meaning "places," is a synonym of the Greek noun *topoi*, the Roman "method of loci" is a technique of memory, not of invention. Moreover, it requires the orator to visualize a very concrete architectural location as an ad hoc container for a verbalizable concept. A Greek *topos*, on the other hand, was an abstract conceptual "place" stocked with a range of standard concepts (Small, 1997:85–87).

Delivery, the actualizing of all the previous four canons, was the performance as viewed from the outside. Here the orator would make full use of the gestural and pantomimic conventions of his culture, those signs that regularly underscore and illustrate the meanings of phrases and clauses in spontaneous discourse. He would also exploit those vocal features of timbre, volume, pitch, and tempo that might best convey his attitude toward himself, his audience, the third-person others upon which an oration is usually centered. Since this canon most closely concerns the public performance of language in other genres, such as song, drama, and rhythmically intoned narrative, I will save these prelinguistic accompaniments to speech for my next chapter, which deals directly with verbal artifacts.

seven
The Poetics of the Verbal Artifact

By tracing the major stages in human prehistory, this book has promised to show how verbal artifacts embody in their structures and themes the cognitive evolution of our relatively recent and, so far, successful species. I began by setting forth the theory of stages proposed by Merlin Donald—namely, the episodic, the mimetic, the mythic, and the theoretic. With language, at the onset of the mythic stage, we entered a new umwelt, the newly imagined universe of discourse, and, as we did, our brains underwent profound changes. As my previous chapters indicate, I agree with those who view this process as gradual. How long it took to form the language-ready brain will probably always be a matter of dispute.

This problem becomes somewhat more manageable, however, if we agree that our modern brain is the outcome of many lines of incremental change, each initiated at different points in evolutionary time, most during the period between the appearance of the first primates (70 mya) and of the first humans (2.5 mya). Each change might be modified by later changes, but in its basic structure it remained to be recruited for further uses. Thus, the wiring associated with episodic and mimetic skills continued to be available to the linguistic mind in the mythic stage.

We could not have survived had we not been able to rely on our primate and prelinguistic resources, but the old and the new mind, housed within the same skull, were not always harmonious housemates. In matters of thought and reasoning, potential conflict was always possible between what has been called System 1 and System 2. The prelinguistic aptitudes of episodic and mimetic consciousness do not always mesh well with language-structured concepts and logic. We find evidence of this conflict in the structures of oral poetry. Its development, which seems to have been in the direction of longer narrative performances, reconstituted extended episodic consciousness within the new medium of language. The nature of oral poetry, not simply as a play behavior, but as a procedure for making separate verbal artifacts, placed these in the category of instrumental products that, like the carefully crafted tools of the mimetic stage, were designed to be saved and reused. The problem was that language, the new System 2 information technology, was a medium not easily adapted to satisfy the episodic and mimetic demands of the System 1 mind.

In this final chapter, I will examine how the preliterate imagination by means of oral poetry struggled to smooth the fault lines between the old and the new mind and devised workarounds when it confronted inherent obstacles. In doing so, it resorted to some of the same stratagems that formal rhetoric would later codify, since extended speeches and extended verbal artifacts face some of the same cognitive constraints. When writing was finally introduced, verbal artifacts made a gradual, partial transition from the public performance venue to the private, two-dimensional page. This radically transformed some, though not all, of the issues involved in the transition from S1 to S2. As literate verbal artifacture took the work of shaping the medium in fresh, new directions, some aspects of the art form were preserved, some radically transformed. In a final epilogue I will suggest that, in those very transformations, writing continued to enhance the age-old power of language to open wide a window on the mind.

The Ritual and Poetic Genres

It has long been assumed that the earliest medium of imaginative expression was chanted speech accompanied by music and dance and that written lyric poetry was distantly derived from this ancient, universal art form. Theories of the origin of poetry tended to stress the emotional and irrational. Rousseau (1754/1984) pictured the individual savage moved by awe or fear to personalize the forces of nature. In a famous statement attributed to him, Jacob Grimm declared its origin to be collective: "The people make poetry" (*Das Volk dichtet*), an idea that influenced the folklorist Francis Child to speak of the "singing, dancing throng" and seek to find traces of spontaneous expressions in the refrains of folk ballads (Child et al., 1904). Friedrich Nietzsche in his celebrated first book, *The Birth of Tragedy Out of the Spirit of Music* (1872/1999), sought the preclassical origins of this poetic genre in a religion that was itself pre-Olympian, an insight that encouraged Cambridge scholars such as James Frazer, Jane Harrison, F. M. Cornford, and Gilbert Murray to explore the prehistoric ritual origins of the performing arts.

Working within the classical tradition, at a time when that was simply called "the Tradition," Murray (1927/1957) identified the Greek *molpê*, a song sung by a harper accompanied by a troupe of dancers, as the antecedent of all verse genres. Since it was such a delight to attend a *molpê*, the ancient Greeks assumed their sky-gods must also enjoy such events. Accordingly, they imagined Apollo as the singing harper and immortal Muses as the chorus that encircled him. "In development, one would conjecture, the group came first and the individual after" (31). That is to say, the "singing, dancing throng" became the *molpê*, which subsequently divided into secondary genres: hymns and choral odes required dancers who were also singers; tragedy and comedy presented chanting dancers together with speaking actors; while epic and song featured soloists who sometimes accompanied themselves on lyres or were accompanied by flute players.

Contemporaneous ethnography had also been supplying essential documents for this search for cultural beginnings. Studies of extant hunting-gathering societies in Africa, North and South America, Australia, and Siberia provided what was deemed suggestive glimpses into Upper Paleolithic and Neolithic folkways and institutions, including the role of song makers and storytellers (Gummere, 1901; L. Pound, 1917). They also provided a number of examples of shorter verbal genres, including charms and riddles, which André Jolles explored in his *Einfache Formen* (1930), a study of the "simple forms" from which longer, complex verbal works were composed, a line of inquiry that Andrew Welsh (1987) was also to adopt.

There is a long scholarly tradition that assumes that Paleolithic cultural artifacts (e.g., cave paintings, glyphs, and figurines) served religious purposes, a religion that has been variously identified with animism, totemism, and shamanism. Whether or not such surviving, ethnographically examined religious practices shed light on the religions practiced as early as 70,000 years ago is difficult to determine, but whatever that practice was, we can be sure that language was a contributing element.

Ritual language would probably have been formal, for talking to gods is indeed "talking to strangers" (Wray and Grace, 2007). But it would also be esoteric and formulaic. Persons outside the community, if they were ever allowed to hear it, would probably find the wording mysterious— and deliberately so, since only the initiated should be allowed to know its significance. That word, "mysterious," comes from the Greek word for initiation, *mysterion*, the root of which means "keeping silent." Initiates were sworn never to divulge the rituals to outsiders, never reveal the words used and their mystical significance. It might be a stretch to assume that, 2,000 years ago, the religions of urban Greeks and Romans resembled those of hunter-gatherers some 70,000–50,000 years earlier. On the other hand, if our concern is not comparative religion, but rather the function of language as an artifactual medium, we might glean some insights from just such a speculative comparison.

The most famous Greek mystery religion held its initiation ceremonies in Eleusis, a center of wheat and barley production, 15 miles northwest of Athens. Celebrated annually near the autumnal equinox, during the planting of winter wheat, the ceremony commemorated the search of Demeter, the goddess of grain, for her daughter Kore, or Persephone, whom Hades, god of the dead, had abducted. Among the few cryptic accounts of the ceremony, we are told it had three features: (1) things done (*drômena*), (2) things shown (*deiknumena*), and (3) things said (*legomena*).

Cross-referencing it to other mystery religions—e.g., the cults of Dionysos, Isis, and Mithras—scholars have pieced together a somewhat fuller view of these three essential features. The *drômena* were actions engaged in by the priests and priestesses of the cult and by its initiands and initiates. These actions included processions, the handling of sacred objects (*hiera*), and participation in liturgical dramas reenacting events in the life of Demeter and her daughter. The *deiknumena* were sacred images and emblematic objects, some of them placed along the walls of the great temple of Eleusis, a structure that by the time of Roman domination had come to fill a space half the size of a modern football field. The *legomena* were, it is supposed, the recitation and mystical interpretation of the myths pertaining to the two goddesses.

The ritual of the Catholic Mass also presents these three features. The *drômena* include the preparations for the Eucharistic meal—the handling first of the paten, or plate, for the bread that will become body of the sacrificed Jesus, then of the chalice for the wine that will become his blood, and finally the distribution of these sacramental *hiera*. The *deiknumena* are the images and emblems placed throughout the church in the form of images—statues, murals, and stained glass. The climactic showing is of the wafer of bread and the chalice of wine. The *legomena* that accompany the *drômena* and the *deiknumena* are prayers that both *effect* the theophany, the transubstantiation of the *hiera* into the sacrificed god, and *interpret* it, using the words of Jesus at the Last Supper: "This is my body. . . . This is the chalice of my blood. . . ." Additional *legomena* usually include readings from Christian Scripture, a sermon, and the singing of hymns.

The Catholic Mass as described here is based on a much more ancient ritual—namely, animal sacrifice—which at some point in prehistory also probably included human sacrifice. Jesus is referred to as the "Lamb of God," a substitutory offering to this deity. When the priest shows the broken wafer of bread, he tells the congregation, "This is the Lamb of God, who takes away the sins of the world," for the slain lamb and the milled wheat are both representations of the slain hero who has offered himself as food for his people. Sacrifice is a complex pattern of behavior that combines a number of deeply embedded human concepts and attitudes formed first in early Paleolithic times with hunting cultures and later modified in Neolithic agrarian cultures, which first appear in the Middle East ca. 10,000 B.C.E. In it we see a mix of themes, from the altruistic ethos of hunter-gather societies, the magical power of blood, and the belief that punished individuals can satisfy the anger of vengeful gods, otherwise directed toward the entire community (Girard, 1972/1977, 1978/1987). The evolutionary sources and meanings of sacrifice are beyond the scope of this book, but, as a ritual paradigm, it has profoundly influenced other rituals and ritualized behaviors and has given the verbal genre of tragedy its distinctive character (Harrison, 1912/1962; Burkert, 1983; Segal, 1999).

Another ritual paradigm, also a universal pattern, is the rite of passage. This marks the passage of an individual, alone or in a cohort of others, from one social identity to another, e.g., from puberty to adulthood. By extension, however, such rites blend into other initiatory ceremonies, weddings, enthronements, and funerals (wakes and burials). Narrative plots, such as we find in folktales and epics, borrow the basic structure of the rite of passage when they represent a hero moving from a familiar to an unfamiliar setting (the condition of "liminality"), where he is challenged by dangerous and even monstrous beings before reestablishing himself, usually at a higher rank (Van Gennep, 1909/1966; V. Turner, 1969). The initiation at Eleusis, in dramatizing the wanderings of Demeter, represent the courageous goddess in a liminal state that the initiands themselves are also forced to enter.

In these ritual paradigms and the particular rituals that take them as models, the three features (*drômena, deiknumena,* and *legomena*) may appear separately, as in wordless action (pantomime), in wordless visual representations, and in direct speech, the latter taking the form of instruction, narration, or prayer. But, as the sacrificial ritual of the Mass and the rite of passage ritual at Eleusis indicate, the three features of ritual are typically presented in parallel, two or three at a time. All three, for example, would overlap when a procession walks or dances, while displaying an image and singing a hymn. There is indeed something of the Wagnerian synthesis of the arts (*Gesamtkunstwerk*) about most public rituals.

While each of these three ritual features appear to correspond to the three sign functions, they each can incorporate more than one function. *Drômena* correspond to indices because ritual actions signify the effects that they are believed to produce. They can, of course, also be iconic when they imitate another action or other actors, as when a priest assumes the person of a god. *Deiknumena* correspond more directly to iconic signs, since they represent the visual aspect of something or someone. On the other hand, the "thing shown" may require the action of a "show-er," which may entail indexical gestures. Of the three, *legomena* seems to correspond wholly to symbolic signs. Yet, as I pointed out in the last chapter, symbolic signs, morphologically arbitrary as they are, have the marvelous capacity to signify indices and icons by enlisting the power of imagination.

Singly or in combination, these three ritual features may help us identify the ur-forms of verbal artifacts because they correspond to universal semiotic principles that are prelinguistic constraints as culturally neutral and binding as mathematics (Deacon, 2003). Semiotically based doing, showing, and saying, therefore, exhaust the possibilities of communicating human meaning—and not only *human,* for as Deacon maintains, they would even constrain the communicative codes of alien intelligences beyond our solar system.[1] The fact that symbols have absorbed the functions of indices and icons through their power to construct mental images also means that language can represent enacted indices and displayed icons to support it in the making of wholly verbal artifacts. Conveyed through an oral medium, a preliterate verbal composition would be associated at every step of the way, from creation and revision to storage and transmission, with visual indices and icons. I will explore this synergy shortly, but before I do so I will propose another avenue of inquiry, a heuristic based on another universal, namely, the topic-comment structure of the sentence.

Linguistics generally assumes that the capacity to learn and speak a language is an innate human trait and that all languages share certain universal structures. On the extent of innateness and on what universals should be included, though linguists differ greatly, most agree that the brain of *Homo sapiens sapiens* was fully language-ready before 100,000 B.P. and possessed complete language skills by 60,000 B.P. when this subspecies migrated out of Africa. Among language universals, most include the topic–comment structure, that dyadic pattern in which a broad, general, known concept is introduced, followed by a narrowly focused view of it. As I noted earlier, some recent theorists have proposed that this structure reflects the much earlier evolved primate skills of manual coordination (Corballis, 2002; Hurford, 2003; Krifka, 2007; Arbib, 2008, 2010).

If, as I am now proposing, the earliest verbal artifacts would have reflected the topic–comment structure of the sentence, we need first to ask what the nature of the ritual topic would be. For the simplest verbal artifacts, e.g., the proverb or the charm, the entire piece might consist of one sentence, so the sentence and the artifact would be structurally isomorphic. As for the more complex, multi-sentence verbal artifacts, e.g., narratives and dramatic speeches, the topic would need to be fully understood in advance. What I am suggesting is that the topical grounding for the paleopoetic artifact would be a familiar, collectively shared human experience, such as a birth, a coming-of-age, a marriage, a death, a successful hunt, a harvest, an enthronement, etc. The earliest poetic genres would therefore correspond to ritual genres that culturally incorporate such life events. Converting the etic into the emic, those rituals would have applied structures borrowed from the universal paradigms to specific circumstances and do so in locally traditional ways.

These ceremonies within a community of 150 persons or fewer (Aiello and Dunbar, 1993) would potentially have involved all members as performers and attendees. Actions would be performed (*drômena*), sacred objects shown (*deiknumena*), and, most important for our inquiry, words said (*legomena*). The familiar occasion and its prescribed arrangement of actions and images would constitute the topic and the words said would constitute the comment. The occasion-as-topic would indicate the ritual genre of the verbal artifact, as being, for example, a birth song, a wedding song, a hymn, a lament, a praise song, etc. Favorite compositions would be re-performed; new ones, if well received, might be preserved and added to the appropriate generic archive. Many of these compositions might be performed independently of their ritual setting, yet still preserve their ritual associations, while some might become assimilated into larger artifacts, e.g., narrative structures, such as epic.

The idea that verbal artifacts began as ritual *legomena* that had become separated from their original settings and yet pointed back to those origins was first proposed by Kenneth Burke in his *Philosophy of Literary Form* (1973). There he spoke of ritual drama, by which he meant large-scale communal enactments, not simpler rituals such as ablutions, healing charms, and the like. Ritual drama, he claimed, was the "hub," the center from which all other verbal genres devolved outward, genres such as epic, tragedy, and, by implication, all the other oral genres, such as elegies, epithalamia, encomia, and epinikia. The literary genres derive from oral poetic genres as spokes from this primordial hub.

Insofar as they seek to impose humanly intelligible meaning on otherwise inexplicable natural events, rituals are persuasive structures of behavior. The words used in such communal actions would therefore have had what we would recognize as rhetorical features. The five canons that Greek rhetoricians devised simply methodized an ancient practice long used to give shape both to rituals and to the verbal artifacts that devolved from ritual *legomena*. The rhetorical model that I outlined in the last chapter appears first of all in nonverbal visual form as the sacred spaces in which ritual dramas take place (Eliade, 1959/1987). In these spaces are separate places (*loci, topoi*) that contain images that are shown to the participants and where they must witness and themselves perform various actions. In ritual, the first two stages—the placing of the significant objects to be shown and the order of their revelation, both determined in advance—correspond to the rhetorical canons of invention and arrangement.

This preliminary pair, common to ritual and to rhetoric, also corresponds to the compositional stages of a verbal artifact, especially in an oral culture. First, a preexistent, ritually related genre is chosen as fitting the current needs of the composer. This genre then dictates the range of subtopics and influences their arrangement. These are the givens, the topical generalities to which the verbal artificer adds the personally derived comments, the unique elements that must catch and hold the attention of his or her eventual audience.[2]

Oral Performance Style

In the rhetoric manuals, ancient and modern, style seems synonymous with refinement, skillful choice of diction, and mastery of figurative language. When viewed functionally, what these traits all help to achieve is the uninterrupted attention of the audience, the sort of control we mean when we say of a speaker that "he held his audience in the palm of his hand." Before I consider how an orally performed verbal artifact is constructed to achieve this goal, I will first differentiate its structure from those we use in other speech situations.

Within a "society of intimates" (Givón, 1979), speech can take the form of short, holophrastic utterances that can be perfectly clear because the speaker is known. The circumstance that frames the utterance is usually also so well understood that its topics are often omitted and only the comments remain. If it is windy and rainy, I might say "'sopen" and point to the window, implying that my addressee, being closer to the window, should close it. If, however, I am "talking to strangers" (Wray and Grace, 2007), I will speak in full sentences (topic + comment, or subject + predicate) and use whatever persuasive means I deem necessary—"Would you mind shutting that window?"

Full, formal speech would also be necessary when the topic is not immediately apparent to my addressee. If, for example, I come into the room on another occasion and find the window open and my laptop missing, I may conclude that the window was opened from the outside and a thief had stolen it. Now an absent third person has to become the focus of my thought, and, when I phone the police, I must describe my situation in full sentences. The open window in the rain storm had led to an *I–You* exchange, which Emile Benveniste (1966/1973) termed "discourse" (*discours*). In the second situation, when I describe the scene and the probable role of that third-person, a person I had seen earlier gazing up at my window, I would be engaging in what Benveniste termed "story" (*histoire*). This story of mine I would not need to embellish with rhetoric, but, should the case reach criminal court, the prosecutor's appeal to the jury would be a story that now might need to be enhanced by rhetorical style. In setting forth the topics (the narrative of probable events), the prosecutor would need to appeal to emotions of vulnerability and justice but, throughout, take pains to avoid any suggestion of pettiness or unfair use of evidence and not seem to patronize the jury or engage in irrelevant levity or banter with any witness. Like the three kinds of utterance just mentioned (the informal, the formal, and the rhetorically enhanced), verbal artifacts are instruments that empower the intentions of their users and in that regard differ from paintings and sculptures, artifacts that exist as objects of visual cognition.

Laying aside short saved and reused verbal forms, and turning our attention to longer verbal artifacts, we can distinguish those that show performers imitating other persons' speech and actions (drama) from those mediated by a single teller. We can then distinguish the artifacts those single tellers present as (1) the imitation of first-person discourse, e.g., love songs and flyting (insulting tirades), from (2) representations of third-person story, e.g., folk tales and epic within which third-person characters express themselves in first-person discourse. Since I have already discussed ritual as performance art, and drama, like ritual, as a form of showing, I will now examine the oral poetics of storytelling with its focus on the actions of third-person others. This latter form of story Aristotle (*Poetics*, 59b33–37) terms "diegetic mimesis," because the narrator not only tells the story as diegesis, but, when characters speak with one another, the narrator must vocally perform their parts in a mimesis similar to that of a stage actor.

As accounts of third-persons' words and deeds, absent in time and space from the circumstances of their telling, oral narratives are stories that must be rendered fully, not abbreviated. Clearly an informal holophrastic style would be unacceptable for such a story. The plain, formal style one might have used in speaking with the police officer would not work either, nor would imitating the prosecutor's rhetorically enhanced summation. This narrative artifact will have to be crafted in such a way that, not only can it be understood when heard for the first time, but it can also be preserved in memory and reused, like any other valuable tool. Unlike the prosecutor's oratory, which is pointedly suited to the circumstances of this one case only, this artifact must also possess a broader significance and a stylistic power even greater than the orator's, because it must be able to hold the attention of re-hearers whenever it is re-performed.

At this point in the cultural evolution of the verbal artifact, we encounter a problem: the brain's capacity to hold in parallel an extended series of data. Episodic consciousness became possible, as Merlin Donald proposed (2001, 2007b), when our primate ancestors, in response to social needs, gradually evolved a more powerful working memory, one that could collate more and more incoming information and execute actions based on that. The human brain further expanded this capacity and developed what he called "intermediate term working memory." As a capacity for processing serial percepts, this was preadaptive to language. But for early humans these serial percepts would have been gestural and vocal and function as indexical and iconic signs. When, as is likely, a protolanguage was developed out of emblematic gestures and holophrastic utterances, symbolic signs enhanced communication and led eventually to the onset of the mythic stage of syntactically composed speech.

As used by language, the auditory channel for serial phonemic differences was, and still is, exceedingly narrow and the flight of "winged words," as Homer called them, exceedingly swift. Working memory may have developed some extended duration for language comprehension, but its limits would have been soon apparent. By reminding the addressee what the context was for each new piece of information, the topic–comment structure of the sentence may have been an early stratagem to keep the serial flow of speech comprehensible. But working memory and its longer form, intermediate-term memory, would not have found a series of sentences easy to follow beyond a certain point without instituting an additional stratagem, a stylistic one also based on the topic–comment model.

An orally presented verbal artifact can challenge an audience's powers of attention, but we might remind ourselves that, in the matter of holding persons' attention, writing also has its own problems. The page, after all, is a very unanimated visual space—no gestures there, no expressive vocalizations to be heard. This helps explain why, while reading, we sometimes find ourselves lost somewhere in a vaguely familiar paragraph, a *déjà-lu* experience that tells us we have been slowly drifting sleepward. But because a book is still open in our lap, we can go back and carefully read that paragraph from the start. Writing thus allows what Walter Ong called "backlooping," but oral utterance has no such advantage: "there is nothing to backloop into outside the mind, for the oral utterance has vanished as soon as it is uttered. The mind must move ahead more slowly, keeping close to the focus of attention much of what it has already dealt with. Redundancy, repetition of the just-said, keeps both speaker and hearer surely on the track" (Ong, 1982:39–40).

These hallmarks of oral poetry have many names and forms, e.g., amplification, *copia*, apposition, parallelism, and the additive style. If that poetry is metrical, these devices may include repeated feet, set numbers of feet within the line, rhymes that link lines, and lines organized into repeated strophes. Metrical or nonmetrical, oral poetry includes phrasal formulas, epithets, and repeated incidents, as well as allusions to traditional verbal artifacts and myths extrinsic to the story being told (Gray, 1971). All these provide the hearer, through anaphoric backlooping, with reminders of earlier acquired information. When competent attendees of an oral performance hear any of these elements repeated, they feel included as participants—they "get it."

Traditional oral narratives seem particularly cross-referential, existing as they do in the cultural context of other traditional narratives. Ancient Greek audiences who heard the angry words of Achilles and Agamemnon (*Iliad* 4) would have been aware of earlier and subsequent events in both men's lives from other sources, including their conversations with Odysseus in the Underworld (*Odyssey* 11). Those other sources were mainly the so-called cyclic epics, which told of the events leading to the Trojan War and its aftermath (e.g., the judgment of Paris, the abduction of Helen, the death of Achilles, the Trojan Horse, the sacking of Troy, and the return of the Greeks). Similarly, the thickly structured mythic allusions in the odes chanted by Greek tragic choruses would have reminded audiences of a shared cultural heritage. Then there is the Anglo-Saxon epic of Beowulf that begins with a cross-reference to the founders of the Danish kingdom, of whose glorious deeds "we have heard."

Such reminders added to the richness of the hearing experience but also helped preliterate societies maintain communal knowledge. Immediate verbatim repetition for the purpose of rote learning is not typical of oral cultures. Instead they use "spaced repetition," a learning method first tested by Hermann Ebbinghaus and subsequently found widely effective. David Rubin has linked this mnemonic technique; with the repeated cross-references and allusions found in oral narratives (Rubin, 1995:24–29; Kramár et al., 2012).

In terms of the topic–comment dyad, every repetition is an anaphoric comment on a preceding topic, a comment that accomplishes two purposes: it primarily reinforces that earlier information and secondarily adds a bit of new information. As a grammatical term, anaphora means a back-reference to an antecedent word or phrase; as a rhetorical term, *it* means the repetition of the same word or phrase at the beginning of a series of grammatical units. (In the preceding sentence, "*it*" refers to "anaphora," an illustration of grammatical anaphora.) These two forms of anaphora are related in that both prepare the hearer/reader for a new piece of information. The actually new forward-directed comment immediately follows the backlooping comment. For example:

And the loftiness of man shall be bowed down [*topic*]
and the haughtiness of men shall be made low [*anaphoric, backlooping comment*]
and the LORD alone shall be exalted in that day. [*new comment*]

Note that this biblical verse contains both rhetorical anaphora ("and . . . and . . . and," admittedly not a very striking example) and grammatical repetition (the parallelism³ of lines 1 and 2), followed by the grammatically and semantically divergent third line. Every repetition is both a return to a topic base and a signal that the speaker is about to launch forth into a new comment, a procedure reflecting the thought process that William James likened to a bird's perching before taking flight (see p. 103).

As we see, repetition can function on the macro-level as allusions to other verbal artifacts and on the micro-level as back-references to previous sentence content. Joseph Russo called repetition the "master-trope of traditional epic phrase-making," a device that might "be conceived in its simplest essence as Item Plus" (1994:374). In the following excerpt (*Odyssey* 4.876–90), he marked three typical forms of repetition in the lines of his translation: appositional phrases, explanatory extensions, and metonymic extensions (the italicized portions represent these stylistic repetitions.) This exchange between Menelaos and the sea nymph Eidothea is part of the story Menelaos tells Odysseus' son Telemachos about his perilous return voyage to his kingdom. Russo points out that there is no "purple patch of rhetoric" in these lines and adds: "Note the many ways in which a word or idea is either repeated or extended, and how certain extensions are tightly bound to the next idea" (375).⁴

"*I shall speak out to you*, for all that you are a goddess,
that it is no way willingly I am held here, but *rather I must have given offense to the gods, / they who keep wide heaven.*
But you now tell to me—*the gods are aware of everything*—
who of immortals fetters me *and binds me from my passage, /*   880
and the homecoming, / how I will make it over the fishy sea?"
So I spoke and she answered at once, *bright among goddesses*:
"Now indeed O stranger *will I speak to you / without guile.*
A certain one frequents these parts, *the unfailing old man of the sea, / immortal Proteus the Aigyptian, / the one who knows /*   885
the ocean's every depth, / Poseidon's underling.
They say he is my father *and that he gave birth to me.*

> If somehow you might be able to lie in ambush *and to seize him,*
> he would be able to tell you the way *and the measures of passage*
> *and the homecoming,* / *how you will make it over the fishy sea."* /   890

The function of repetition as a means for hearers to process an extended series of words smoothly and without loss also conforms to Roman Jakobson's definition of the poetic function as "the projection of the principle of equivalence from the axis of selection to the axis of combination" (1960:358). Repetitions of the sort we are here examining are, of course, as-if equivalences, as indeed the second term (the "source") of a metaphor is also an as-if equivalent of the first term (the "target"). When a repetition is inserted into the axis of combination, it creates an instance of parallel cognition that momentarily slows the serial march of syntactical units that might otherwise overwhelm the working memory capacity of an audience. This is especially necessary because oral discourse and narration tends naturally toward *parataxis,* that straight-ahead series of simple sentences and information units connected only by words such as "and" and "then." Repetition in oral poetry may be classified as paratactic, but only if we understand this as a backlooping form, which we might term "anaphoric parataxis." (As for *hypotaxis,* the use of subordinate clauses, this stylistic practice would have to wait until writing was introduced.)

The power of speech, as a symbolic system, to absorb the functions of earlier communicative codes, based as they were on perceived indices and icons, seems to have provoked the ingenuity of *Homo sapiens sapiens* to recover within language the multimodal richness of the *pre*symbolic world. Aristotle had noted the rhetorical advantage of placing images "before the eyes" of one's audience. This primary power of language to prompt the imagination is indeed considerable. To account for this working relation Allan Paivio (1990) developed his "Dual Coding Theory" positing two independent representational systems: one was imagery, the analog representation of predominantly visual sensory experience, while the other was language, the symbolic representation of experience as words in sentences. Images, he said, represent objects in part–whole relationships and parallel-processes them, whereas language represents them in hierarchical relationship and processes them serially. The connection between these two codes is reciprocal: a word will cue an image and an image will cue a word. This conforms to the dyadic pattern with the imaginal code broadly representing spatially situated figures-in-grounds and the verbal code narrowly representing a temporal stream of phonemes and morphemes. This is also consistent with the claims of Dual-Systems Theory, if we understand imaging as a function of prelinguistic S1 thought and words as a function of later, language-based S2 thought. We, as inheritors of the latter, more recent system, have also inherited the earlier system and hence our thought is both analog *and* symbolic, imaginal *and* verbal.[5]

Memory

Mental mapping and personal retrospection, both stored in a format shaped by figure–ground principles, as excursions into the there and then, were departures from the here and now. Early *Homo*, like other animals, was predominantly a creature of the Present, but unlike them he had an ever-increasing capacity and need to detach himself voluntarily from the here and now, to reflect upon already lived events and to plan new ones. If we may speak of prelinguistic tenses, retrospected experience provided our early ancestors with a new tense, the Past Absent, as a cognitive alternative to the Present, while the mental mapping of a territory provided a preliminary opening into the Future Absent, a temporal realm that would open wider in the Middle and Late (or Upper) Paleolithic eras. With language, an efficient means was devised by which both these two absent realms could be opened up for mental travel.

Style in oral poetics takes what Merlin Donald has characterized as the extended consciousness of the episodic stage and adapts it to the specific nature of verbal communication. What classical rhetoric termed "memory" was an adaptation of the mimetic stage to other language-associated needs. According to Donald's chronology, it was during this first age of human technology, a period of over 2 million years, that our ancestors developed those habits of planning, collaboration, and executing that resulted in tool making and the use of tools to make other products, such as shelters, hearths, garments, and ornaments. The serial process by which productive actions can be imitated, a prototype reproduced, or an idea externalized as a completed product became the model for the production of verbal artifacts at the inception of the mythic stage. These productive skills all involved learning (i.e., committing to memory) sets of serial actions.

In rhetoric, the function of style was to facilitate the extension of an audience's working (short-term) memory *during* the performance of a speech, whereas the function of memory was to store it *between* its composition and its delivery. Just as the rhetorical canon of style has its equivalent in oral poetics, so also does the canon of memory. The latter requires a long-term storage system. But which? Does the brain have a special neural wiring system for extended sequences of sentences? Certain rare individuals, like Luria's (1987) mnemonist, have this capacity, but thankfully few of us are afflicted by this gift. Semantic memory, a long-term system, includes our capacity to remember words and their meaning, but it does not seem well suited to the laborious task of storing long sequences of words and incidents. The specific memory system recruited for this task is episodic, or autobiographical, memory, a store that is often imagined as a spatial entity, a repository with compartments that we rummage through to view their contents. This, as I noted in earlier chapters, catalogues events using time and space coordinates. After meditating on the powers of perception, Augustine in the tenth book of his *Confessions* speaks of entering the "grounds and spacious palace of my memory wherein lie storerooms of innumerable images, brought there from all sorts of sensed objects." To retrieve an account of a personal experience from episodic memory, we need to visualize its setting and project a rough sequence of actions. This contrasts with semantic memory, our long-term store of beliefs, facts, concepts, words, and image schemas (Tulving, 1983).

When we employ the term "episode," we must distinguish between Donald's and Tulving's usage. When Donald uses the word, he usually applies it to a "perception event," a series of actions happening here and now. As the episodic stage proceeded, our prehuman primate ancestors developed the capacity to process a rich multitude of details during an ongoing social interaction, or episode, through an extension of short-term working memory. When Tulving uses the "episode," he refers to a retrievable experience stored in long-term memory. Both "episodes" are unitized actions, one in the present, the other in the past, but the latter, when it is retrieved, is re-presented in the mind of the person who once experienced it. When it is told to others, it escapes from the personal past and is reincarnated in a new social present as a story.

We should also be clear what we mean when we use the word "memory." When in English we say we "have a memory" of X or Y, we refer to a personal experience stored in episodic memory and not to the system itself or to the faculty in general. Personal episodic recollections ("memories") seem to possess a substantive, three-dimensionality, unlike our recollection of facts, beliefs, and the meanings of words, which are stored in the general knowledge archive of semantic memory and have nothing personal about them. Yet, as I observed in chapter 4, we rely on semantic memory when retrieving personal episodic memories because the semantic system has direct access to image and motor schemas, those indispensable elements of any recollected episode.

Lived episodes, as they are happening, are quite complex affairs. Occurring in a particular setting, they may involve multiple senses, including motor sensations. Episodes may have verbal elements, but even these have multimodal accompaniments, e.g., tones of voice, gestures, joint gaze, and pointing. Such episodes may also involve multiple agents in overlapping activities. These we process in parallel or serial mode with shifts of focal attention that reflect their relative salience.

The complexity of such online episodes poses major problems for retrieval. The more emotionally arousing the experience, the more likely its details will be stored in long-term, episodic memory, but an episode rarely floods back to consciousness in all its simultaneous details and full emotional force. An online episode has its own identity, though it is still part of a temporal continuum. A retrieved episode, on the other hand, once it is extracted from storage, has a definite beginning and an end, each bordered by a dim fringe of forgetfulness. When we mentally reconstruct the event we use semantic memory, our knowledge of the world, to suggest to us how things must have happened. (The tendency of such reconstructions to supply stereotypes makes eye-witness testimony subject to distortions.) In retrospection we also tend to serialize what may have been originally perceived as parallel and this sequencing may lead us to suppose a cause-and-effect relationship where none may have actually existed. As Tulving has noted, episodic recall is effortful and the reconstructive process accounts for that.[6] When we choose to tell others this story and transpose it to the medium of language, we process it even further. What in our private recollection was only a tendency to serialize is now a rule firmly enforced. We may want to convey the multiple simultaneous actions of multiple agents, but our success is limited by our medium and by the capacity of our hearers to convert paratactic sequential input into parallel imagery. We may say, "While X was doing A, Y was doing B," but, if we want to add how halfway through Y's action, Z began doing C, our hearers and we will find ourselves juggling just too many balls in the air at once.

Like the Past Absent and the Future Absent, there was yet another mental realm that early humans would have been able to enter. It, too, was episodic in structure and would eventually provide an important model for verbal artifacts. In it, scenes appeared, faded, and were replaced by new ones. In some respects it resembled retrospected experience, since familiar persons and places could be reencountered, but unlike retrospection it could not be voluntarily accessed. This realm was dream. Every night some part of the sleeper, whose body remained where it lay, seemed to travel elsewhere and act like a child or do violent acts or fly like a bird, while the dead walked upright and the trees and rocks transformed themselves into animals. If a mentally mapped terrain and a remembered event were real, though absent, so too the dream world might be real, though absent. But was it actually absent? After all, when awake, one also confronted unanticipated, unbidden events. Perhaps when one fell asleep one reawakened in another world, a strange and perilous world, but one in which one's beloved dead might be restored. In a grammar of mental states the tense in which one wandered in this paradoxical realm might be called the Absent Present.

Our difficulties remembering our own past episodes and then telling them to others is compounded when that episode is a dream. While dreaming, we normally accept as real every incident, every shift of scene, and we cannot halt the sequence of events to question or voluntarily explore a setting. Here persons and things can have multiple identities and change shape as readily as Eidothea's father, Proteus, the Old Man of the Sea. As we awaken, we are sometimes so moved by a dream that we think that by recovering its sequence of incidents we can reexperience its power. This usually fails. The aura of the dream dissipates quickly. We are left with a broken chain of merely odd happenings. Then if we try to tell others, we succeed only in boring them.

The retrieval of actual and dreamed episodes forms the basis of narrative. But constructing survivable verbal artifacts is no simple matter. The "I-did-this-then-I-did that" style could never work in a preliterate culture. I have spent this much time examining episodic memory in terms of its cognitive constraints because I believe the success of verbal artifacts has largely depended on finding ways within and around these constraints. The structure of oral narrative in particular may be usefully analyzed as a set of strategies aimed at crafting collectively accessible episodic memory content.

The five rhetorical canons, of which memory is the fourth, were certainly not the source of oral poetic practice, which must have been as old, or nearly as old, as language itself. The canons were, I suggest, a set of well-proven principles based on many millennia of skillful storytelling, newly adapted to a particular civic culture. Just as Newton did not invent gravity in 1687, but codified it as a natural law, Aristotle did not invent rhetoric, much less the cognitive systems that rhetoric—and poetics—depend on.

"Memory," as a rhetorical canon, is the means by which an orator commits a speech to long-term memory. Having located all the parts of his argument in "common places," he relocates these found tools in mnemonic *loci*. At least until he delivers it, he must remember the topics he has found, the order of their deployment, and the stylistic features he has chosen to make them most effective. The length of storage time extends from the moment he has put the final elements of the speech together, through the time he has rehearsed it and committed it to memory, to the moment he stands before his audience and delivers it. Once uttered, the speech is over. If literate, he can, of course, save his notes or the total transcript, but the speech, as such, is a one-time affair. The oration is itself a complex found tool, an ad hoc artifact. It may be a finely wrought instrument of persuasion, but it is as evanescent as a conversation.

Memory, as a factor in oral poetics, is also long-term, but the artifacts it stores have no expiration dates. This memory, moreover, is not confined to the brain of the verbal artificer but belongs to a transgenerational, collective archive. Being dispersed within a community of hearers, a verbal artifact will be remembered differently, so variants of the initial composition will coexist and potentially vie for the attention of audiences. Both performers and audiences have a stake in the effectiveness of a verbal artifact. Performers are rewarded for the degree to which they please audiences, and audiences determine over time which artifacts and, among them, which variants deserve to be preserved. While it is unlikely that "the people make poetry," it is certain that within an oral culture they have always had a major hand in *editing* poetry.

Just as mnemonic technique was a key factor in the temporary storage of an oration, it had always been crucial to the preservation of oral poetry. The "mental walk," which I described earlier (p. 174), entailed assigning topics to images, often strikingly bizarre, then placing them in a particular order around a familiar setting, e.g., a street, or the interior of a house or temple. Having done so, the orator, as he spoke, could proceed from topic to topic, visualizing each associated image in turn as he mentally strolled through this location. For the storyteller of a traditional narrative, the mental walk is often represented by the physical journey of a third-person character from one extraordinary adventure to another. The word "adventure" originally meant an arrival, and in many an oral tale an episode commences when a hero arrives at a new *locus* and confronts a new challenge that often takes the form of a *monstrum*. This is literally an indicator, or omen, which may come through a supernatural agent, e.g., a god, a ghost, a talking animal, or a "monster." The *monstrum* serves as a landmark that points the hero in the direction of another landmark, another *monstrum* with whom he has another arrival-adventure. We see this episodically linked structure in the Epic of Gilgamesh, the *Odyssey*, the *Ramayana*, the Book of Exodus, *Beowulf*, and the Grail legends, as well as in the separate Greek myths of Herakles, Dionysos, Jason, Theseus, and Demeter. The resemblance of such narratives to rites of passage and pilgrimage is also suggestive.

Presumably based on oral sources, these quest narratives became models for later literary narratives (e.g., the Greek Romances, the *Aeneid*, *The Divine Comedy*, *Orlando Furioso*, *Don Quixote*, *Pilgrim's Progress*, and *Gulliver's Travels*), as well as tales in the picaresque tradition. The list of such narratives, oral and literate, is so vast that only some universal cognitive predisposition can rightly account for it.

We know that for verbal information to survive in an oral culture it must be retold. This means it has to remain in the memory of the hearer long enough to be communicated to others and interesting enough to them to warrant a retelling. Repeated phrases and rhythms are useful mnemonic devices, but to be stored in the communal memory, the content of an oral composition has to be extraordinary enough to stand out against the background of routinized village life and compete successfully with other oral texts for a niche in the communal memory. It would follow, then, that if two variants of the same story are current, one in which the hero escapes from a giant by slipping away under cover of darkness, the other in which he escapes by donning a helmet that renders him invisible, the latter variant might stand a better chance of surviving longer in oral transmission. It would logically follow that the more counterintuitive details there are in an oral narration, the more memorable it becomes. But this is not necessarily the case. If it were, our dreams would be eminently memorable, for not only are they filled with unpredictable bizarrerie, they seem based in a "reality" with no ontological stability.

In an oral culture, verbal artifacts cannot survive simply by packing themselves with strange and miraculous incidents. Instead, they must reflect the consensual reality of their audience with just a minimal amount of counterintuitive details (Boyer and Ramble, 2001; Norenzayan et al., 2006). As a journey is not "about" the landmarks that guide the traveler, so also an orally transmitted verbal artifact is not so much "about" the extraordinary incidents that occur along the way as it is about the character of the adventurer, his or her resourcefulness, moral decisions, and ultimate goal. As I have suggested, rhetorical and poetic mnemonics share some of the same cognitive processes. The relation between a counterintuitive feature and the everyday representation of reality that accompanies it resembles the relation between the bizarre image that an orator associates with a portion of his argument, places as an imaginary landmark in a remembered landscape, and then visualizes as he mentally walks past it during his speech. Like a successful oration, an oral artifact must ground itself in the consensual reality of its audience. The one important difference is that, while the spoken oration never includes the images the orator uses as mnemonic cues, the orally transmitted narrative always does include them. The reason it does so is that it must survive by self-replication *in* the memory of *others* and its identity as this or that narrative is consequently forever linked with whatever mnemonic scaffolding has most successfully preserved it.

If, as I suggest, the extraordinary nature of its details is one important means of its survival, what we call the "content" of a myth may be impossible to disentangle from its other built-in mnemonic devices, such as formulas and repetition. As paper, ink, and typeface are elements necessary to the survival of information in a print culture, magic, gods, and all variety of *monstra* are necessary to its survival in an oral culture. If, then, the oral message is so dependent on the oral medium, can there ever be a message distinct from this medium, a factual content with truth-value that is extricable from the necessary format in which it is conveyed? If the Greek word for truth (*alêtheia*) has any bearing on this discussion, the question of truth embedded in the husk of fable is meaningless in an oral culture, for, in the absence of the critical habits that literacy promotes, the "true" is simply the unforgettable (a = not + *lêth-* = forgetting). What we generally refer to as "mythology" is thus the genre that memory-dependent oral transmission naturally engenders.

A series of merely extraordinary visual events, however, would be as forgettable as a dream quickly becomes upon waking unless those events possess the sequential form of narrative, a feature that is both mnemonic and familiarizing. This form is mnemonic in that it imposes a before-and-after, cause-and-effect structure on these items of information. It is familiarizing in that it uses the linear format of everyday speech and memory retrieval, thereby transmitting a series of episodes as though they had been somehow experienced or witnessed by the narrator.

Extraordinary events are necessary, but not sufficient for the survival of a verbal artifact in an oral culture. To be preserved in communal memory it must also reinforce communal beliefs. That is to say, the early books of the Bible, the Homeric epics, the German *Märchen*, the English folk ballads, the tall tales from the American frontier, contemporary "urban legends," and every other oral artifact must have been shaped by tacit cultural assumptions and have survived only because people found them therefore meaningful and enjoyed repeating them. When we find that a certain oral text survives over time, we may conclude that forgettable and cognitively dissonant portions were edited out and that which remained was consistent with the moral, political, and aesthetic preferences of the community.

Enacting the Verbal Artifact

Corresponding to the final rhetorical canon, "delivery" (L. *actio*, Gr. *hupokrisis*), oral performance is the process of actualizing the verbal artifact. At this point the performer's cognitive skills of style and memory must be displayed through his or her mastery of the motor skills of voice and gesture.

The verbal artifact, which now unfolds itself in the span of its actualization, is, like any instrument, a means of extending certain human powers. In an oral culture the principal power that narrative enhances is the capacity to time-travel by using the structures of episodic memory, which in this case is used to revisit a past that those now living have never personally experienced. Richard Schechner (1985), speaking of ritual performances, referred to two effects. One was *transformation*, the life-altering result of, for example, an initiatory ceremony. The other was *transportation*, the temporary experience of being elsewhere and other. The instrumental use of a verbal artifact, detached from ritual, produces, at best, the latter effect and a performer's status is judged by how well he induces this state in an audience.

The Greek term *diêgêsis*, usually defined as narrative, is a noun formed by combining *dia* (through) with the verb *hêgeisthai* (to go ahead of, to lead). Diegesis therefore implies the conducting of an audience through a series of events, enacted in particular places and involving third-person others. A narrative thus creates a fictive journey, a movement of consciousness through a spatiotemporal environment, the audience imagining themselves to be traveling toward objects that gradually grow in size and detail or watching as these objects loom in the distance, approach, and then pass by. Persons appear. They speak to one another with heightened emotion and express their feelings in action. Gaps in space and time divide the narrative into episodes: in a few words a new place is announced, a place of new arrival and adventure that is now suddenly enfolded in light.

This process mirrors the discovery of the contents of "common places," the process of invention. It also mirrors the "method of *loci*" by which orators would memorize a sequence of arguments, though, of course, a narrative is not a sequence of arguments but one of incidents and episodes. While it shares many of the stylistic devises of oratory and often portrays persons intent on using words to persuade others, its play frame obviates the need to persuade an audience as to this or that present line of action. As Goethe said of epic, its subject is the "completed past" (*vollkommene Vergangenheit*). The performer and audience engage in fictive play when they agree to believe this past is being brought to life again, but this play is not necessarily the same as fiction. In a traditional oral culture this communal past is also the unforgotten, which makes it ipso facto *alêtheia*, the truth. The test for the performer, then, is the degree to which he brings the members of the audience with him—transports them—to this completely past, completely true, completely other world.

The communal memories awakened in the narrated episodes transport the audience to a world that is indeed other than the world they now inhabit, for in this ancestral world the men and women are taller, stronger, wiser, and more beautiful than their descendants. When they are not themselves gods or children of gods, these persons are intimates of divine beings, who counsel them, fight alongside them, and save them from peril. The selective pressures on orally transmitted narratives ensure that they are interspersed with miracles and *monstra* and that the highest cultural ideals of the community are embodied in the actions of their heroes. The assumption that this sorry world has degenerated from some Golden or Heroic Age seems common to all traditional oral cultures.

Oral performance typically involves special vocal techniques, such as alterations in pitch and amplitude. Tonal levels may be strikingly more varied than in conversational speech, and, if the audience is large, the voice must be loud enough to reach all hearers, a rise in amplitude that also implies a rise in emotional intensity. It also will tend to be more rhythmical. A measured, backgrounded repetition of phonatory features (e.g., pitch, duration, and amplitude) induces subtle entrainment effects in listeners, synchronizing breathing, pulse, and possibly brainwaves. It also enhances working memory (Merker et al., 2009). We naturally associate accelerated heart beat and breathing with an increase in quantity and speed of information processing. During the performance of a high-arousal episode, rapidly repeated movements and sounds can therefore induce in viewers a heightened state of consciousness.

If, in addition to the solo voice, rhythmical dance movements, gestures, and percussive sounds are included, these accompaniments can intensify the reception of the narrative. Ululation, clapping, drum beats, whirling, stamping, all these stylized actions use as their formula *repetition = intensity* and serve to represent the expanded input of the working memory as it confronts an arousing, perhaps overwhelming perception event. Repetition, as I noted in my discussion of verbal style, is a way to stimulate the two forms of episodic consciousness, the online experience of the present and the offline recollection of the past. By the speed of their pulses, these paralinguistic elements may also serve to represent the sort of intense experience of episodic awareness that can extend short-term working memory into an "intermediate term" of lengthened duration (Donald, 2001) and, by its regularity, prepare participants to contemplate images of the past, using as their model the long-term episodic memory format (Tulving, 1983).

The episodic aspects of these two modes—on the one hand, a rapid rhythmic repetition of present movements and sounds, and, on the other hand, a slower, visual re-presentation of the past events—correspond rather closely to Nietzsche's well-known distinction in *The Birth of Tragedy* between the Dionysian and Apollinian factors at the heart of Greek tragedy and modern music drama. These two gods represented for him the polar oppositions of selflessness and self, rapture and reason, dance and dream, music and vision. At the Dionysian pole, this sense of separate individuality surrenders to a mystical participation in a larger reality experienced within an extended episodic present, such as Donald has described. At the Apollinian pole, we enjoy an illusory, yet stabilizing, belief in the unique wholeness of our self, a belief grounded in the *principium individuationis* (the principle of individuation). The latter, as he understood from Schopenhauer, was formed by our sense of existing in time and space—in particular moments and particular places. Episodic memory, which Tulving has maintained is the basis of "autonoesis," is our knowledge of our self as a separate identity. It is this memory system that assures one that, every morning one wakes up as the identical self that last night went to sleep, that over a lifetime of many decades one is that same person who once was a child—in short, that one continues to be *one*. This is also the basis of Nietzsche's Apollinian principle.

Why Nietzsche's polarity has seemed so intuitively compelling is not because of what it says about Greek culture or the birth of tragedy, so much as what it says about the way we frame our experiences and situate ourselves within those frames. When immersed in our dealings with things and persons, we forget our autobiographical selves, as a constantly updated mix of multisensory information flows about us. This, for Nietzsche, is the Dionysian pole, a spatiotemporal field suffused with music, dancing, masking, and metamorphosis. The Apollinian pole, which Nietzsche associated with dreaming, represents the individuated self, a subject contemplating a world of visual objects.

Nietzsche's full title, *The Birth of Tragedy Out of the Spirit of Music*, reminds us that his principal aim was to show how Greek tragedy united music and words and, with them, action and contemplation, chaos and order, reality and dream. Nevertheless, among his many insights there are visual implications worth exploring. When we are involved in visual action, mediated by the dorsal stream, we place ourselves within a three-dimensional spatial field and calculate the position and movement of objects relative to our position and movement. Despite its name, within this egocentric frame of reference we tend to lose our sense of self as we monitor instead the continuously changing optic flow (Milner and Goodale, 1995). This then corresponds to the Dionysian pole. Alternatively, when we step back and contemplate a visual array as objects in figure–ground relation to one another, we place them in an allocentric frame of reference, mediated by the ventral stream. This perspective corresponds to the Apollinian pole.

In evolutionary terms, Dionysos represents the prelinguistic mind, whereas Apollo represents a mind tempered by speech and reason. While the dyadic union of the two through artistic performance would coordinate what Dual-Process Theory calls System 1 and System 2, such evolutionary crossovers are seldom problem-free. One of the problems in the oral performance of verbal artifacts is the persistent competition between the here-and-now of the visuomotor performance and the then-and-there of word-cued mental images. This is an instance of visual interference, as perception and imagination struggle for attention. When music is part of a verbal performance, an auditory interference is likely, as musical patterns and amplitude compete with the phonemic and tonal features of speech. The relation of music to words is problematic, as Richard Wagner, Nietzsche's then culture hero, knew all too well. How does one energize a line of verse with musical power and still render its words intelligible to one's audience?[7]

The power of phonation to communicate emotion testifies to the primate past we all still carry within us. As perhaps our oldest communicative medium, voice resonates deeply within us even when, in normal conversational speech, it functions paralinguistically as intonation contours and grammatical stress. Narrative could never afford, however, to ignore the possibilities of voice, its tonal registers and dynamic variety. In the enactment of verbal artifacts, structures would have to be devised to balance the relative prominence of those two features of speech, phonatory sound and articulatory meaning.

Paralanguage, Protolanguage, and Oral Poetics

The prelinguistic communicative codes became peripheral to speech, but they never vanished altogether. Vocalization and gesture survived as paralanguage, and, because they helped harmonize the old brain with the new, they had important functions to perform, which we can trace in the cultural evolution of verbal artifacts.

The issue of vocal paralanguage, as distinct from linguistic signs, brings us to the topic of prosody, which linguists understand as the use of pitch, amplitude, and duration in the act of speaking. Each element is perceived contrastively, pitch ranging from high to low, amplitude from loud to soft, and duration from long to short. These are said to be *suprasegmental* features in that they are distinguishable from segments (phonemes, syllables, and words) and supply these lexical elements with an overarching character or affect. (Cf. the rising intonation contour of most English questions and the falling contour of most declarative sentences.) As a poetic property, prosody is associated with meter, a suprasegmental pattern that provides poetic discourse with pulse-like rhythms of expectation variously marked at the syllabic, verbal, and phrasal levels.

The anatomical structures that support prosody began to appear 300 mya, when vertebrates first heaved themselves out of the sea and evolved an early model of the air-breathing respiratory system. From our own relatively recent evolutionary perspective, we can consider prosody as the means our highly social primate ancestors used to communicate their needs and fears. For the earliest humans of the Lower Paleolithic, vocalization continued to be a useful way to attract the attention of kin and companions in order to influence their behavior.

What is essential to note is that prosody, as a linguistic phenomenon, is phonation, *not* articulation (see p. 131). Though its features can be somewhat modified by altering the shape of the oral cavity and mouth, prosodic sound is itself a function of the diaphragm, chest muscles, and larynx, the latter being the final determiner. The relative tension of the laryngeal, or vocal, folds determine the pitch of any phonatory sound, the amount of air and the force with which it is expelled through them determines its volume, and the control with which these operations are performed determines its duration. Humming provides a simple demonstration of phonatory mechanics. When we hum, we may do so either with closed mouth, letting the sound resonate nasally, or with an open mouth ("oral phonation"). In either case, the tongue makes no movements or in any way engages the palate, teeth, or lips: the vibrating vocal folds control the action entirely. By contrast, when we whistle, our vocal folds are held open as the air from the lungs rises into the mouth, where the tongue regulates the size of the oral cavity and the lips regulate the size of orifice through which the air exits. The mouth, in effect, employs articulatory mechanics to mimic the phonatory action of the vocal folds.

We open wide our mouths and phonate when we are surprised, endangered, or enraged or when we experience overwhelming pleasure, pain, or sorrow. In such circumstances, the sounds we make range from extended vowels and diphthongs (howls, growls, and groans) to loud, spasmodic glottal exhalations (laughter and sobbing) to softly extended exhalations (sighs). Such is the expressive repertoire, inherited from some 55 million years of primate experience, that, now integrated into human speech, resonates in the background of articulated words as prosodic paralanguage.

Humans became fully articulate when they developed the lingual control necessary to engage the other "articulators"—soft palate, teeth, and lips—and thereby produce a variety of distinct phonemic consonants. Whispering demonstrates how oral articulation can be detached from laryngeal phonation and as such is the diametric opposite to humming. Whisper a set of words. (Note, by the way, that you cannot whisper anything that does not sound phonemic and "word-like"—try whispering the sound of a creaking door, an owl's hoot, or a horse's whinny.) While whispering, as in whistling, the vocal folds are held open, but now the tongue is fully engaged with the other articulators: since the larynx is not producing, and the mouth cannot produce, pitch variations, the whisper is monotonal. Its volume remaining relatively uniform, the surest way we have to emphasize a word is by lengthening its duration.

In natural speech, prosodic features constitute the vocal/auditory ground, while phonemic articulation produces the focalized sound figures. We *hear* the prosodic features while we *listen to* the words. The employment of linguistic prosody in the constructing of verbal artifacts, not only oral, but also written, further indicates its ongoing relevance to human communication. It should be noted, however, that compositions, such as narrative, that require close attention to episodic details keep prosodic structures well in the background. The word "epic" derives from the Greek word *epos*, speech, a fact that underscores the importance of the narrator's and the third-persons' words are to hearers of a narrative performance. Speech, especially urgently uttered speech, is prosodically irregular, its patterns unpredictable. The regularity of epic meter, therefore, is varied to accommodate speech rhythms and operates in the periphery as a calming phonatory accompaniment to the voices of agents and images of action that words jaggedly evoke.

Just as prelinguistic voice in the form of prosody continued alongside language, prelinguistic gesture also continued. As David McNeill and others have amply demonstrated, gesture, as an expressive accompaniment to speech, is a universal human behavior, a fact that strongly supports the notion that it served that same purpose prior to the earliest migrations of *Homo sapiens sapiens* out of Africa (100,000–70,000 B.P.). Before we can explore these, however, we should first consider what kinds of speech-associated behavior were—and still are—mediated by visible displays, conveyed not only manually but also facially and posturally. We should also note that they are not restricted to the musculature and integument of the body but may also take the form of clothing and ornament. In brief, we may refer to these indexical gestures as *intrinsic* (made *by* the body) and *extrinsic* (worn *on* the body).

In a society in which a gestural system of communication has been effectively superseded by a vocal system, speakers would still need to draw attention to themselves and, through a variety of indexical signs, convey affective information to supplement their symbolic (holistic or segmented) utterances. These would have included a set of manual and brachial movements not dissimilar to those we still use in conversation, but these would also include orofacial indices, e.g., scowls, smiles, bared teeth, stares, and eyebrow flashes, as well as head movements and postural stances. To enhance this kinesic paralanguage, men and women would draw attention to these body parts by applying extrinsic indices in the form of paint, ornamentation, garments, and other accoutrements. The eyes, brows, and mouth would be accentuated so that their affective gestures could be seen at a distance. Necklaces could also be used to frame the face in a draped semicircle that mirrored from below the upper semicircular outline of the skull. Bracelets and anklets could also make the body more expressive at a distance.

Unless performed in total darkness, spoken art forms always focused the visual attention of the audience upon the speaker. To reinforce this attention, this person may have been specially adorned, garmented, or masked. In most instances, the performer would be conveying a traditional composition and would be assumed to be transmitting the words of another person (cf. *imitative play*, pp. 72–75). The strong visual presence of any performer tends to merge that person with the image of the person he or she represents. The performance would therefore have the character of a self-transcending act, a resuscitation of the dead perhaps, or a theophanic vision. The total-body percept projected through kinesic paralanguage would therefore constitute a complex iconic sign that would tend to veil, if not wholly occlude, any mental images that might otherwise be evoked by the spoken words. This is the interference effect I mentioned earlier. Unlike the words one reads from a printed page, the spoken words of an oral poem are not addressed exclusively to the imagination; instead, the costumes, *mise-en-scène*, and the performer's delivery constitute mental imagery externally actualized as spectacle.[8] The oral performance of verbal artifacts thus derives much of its power to move audiences from exploiting earlier, more deeply embedded modes of primate communication, prelinguistic forms that persist in spoken discourse.

In addition to phonatory and gestural elements, oral performance exhibits elements that may be identified as protolinguistic—namely, holophrastic utterance. Here I refer, of course, to Alison Wray's theory of language origins (see p. 117), centering on the "formulaic sequence," defined as "a sequence, continuous or discontinuous, of words or other meaning elements, which is, or appears to be, prefabricated; that is, stored and retrieved whole from memory at the time of use, rather than being subject to generation or analysis by the language grammar" (Wray and Perkins, 2000:1). This form that Wray has proposed as the protolanguage that directly preceded full, compositional language (1998, 2002b) has obvious implications for the study of orally composed and performed epic, seeming to fit quite well into the Parry-Lord oral-formulaic theory. The latter sought to account for the ability of performers of epic to improvise their performance by inserting formulaic phrases in places appropriate to the narrative and to places in the metrical line. Milman Parry and his student, Alfred Lord, having discovered this practice among Balkan epic singers (*guslari*), reexamined the Homeric epics and found evidence there that the same method of improvisation had been used.

Other scholars subsequently found clear traces of formulaic language in Beowulf and other oral-based narratives.[9]

Oral-formulaic theory has focused on how helpful formulas are to performers. I suspect, though, that they would not have been so frequently used if they were not also helpful to audiences. An audience enjoys the fluent telling of a tale and becomes uncomfortable when the teller seems to struggle to remember what comes next, a gap that a familiar formula may be used to fill. A performer who thus seems to be a spontaneously overflowing fountain of words will have a better chance of holding an audience's attention.

Another reason audiences find formulas helpful will seem rather counterintuitive: formulas sound spontaneous. When we have a lot to tell someone, when we "speak from the heart," we make no special effort to compose a string of words never before uttered. Spontaneous speech, when analyzed, is found to contain an ample amount of familiar idioms, homely prefabs, if you will (MacKenzie, 2000; Lin, 2010). Verbal artifacts, stored collectively and re-performed at semiregular intervals have formal features, to be sure, epithets and other archaic phrases that are unlikely to be used in conversation. Nevertheless, this formulaic diction, when culturally familiar to an audience, is recognized as a kind of spontaneous, idiomatic speech. We note this, for example, in the ever-available biblical phrases that fill the improvised sermons of skilled preachers (Rosenberg, 1970).

Formulaic speech in oral performance is another instance of anaphoric repetition. On its function, the Swiss medievalist Paul Zumthor wrote: "Rather than as a type of organization, the formulaic style can be described as a discursive and intertextual strategy: it inserts and integrates into the unfolding discourse linguistic and rhythmic fragments borrowed from other preexisting messages that in principle belong to the same genre, *sending the listener back to a familiar semantic universe* by making the fragments functional within their exposition." (Zumthor, 1990:89, italics added).

As anaphoric repetitions, these vestiges of an ancient protolanguage thus function as music-like leitmotifs. Among the "design features" of music, William Tecumseh Fitch (2006) lists *performative contexts* (music is associated with certain social gatherings) and *repeatability* (musical pieces may be heard over and over again). This, he says, differentiates them from most utterances, which, once uttered, are never consciously repeated. There are, however, recurrent circumstances that call for recurrent utterances. Among these are greeting and farewells, prayers, traditional stories, and dramas. These ritualized discourses are either whole formulas or abound in formulaic language (Fitch here cites Wray, 2002a). They also show another interesting trait: "such formulaic utterances have often been singled out by linguists as peculiar [because] their very similarity to music seems to differentiate them from ordinary language" (Fitch, 2006:180; see also Mithen, 2006:12).

Fitch's claim that music is socially contextualized and repeatable and that certain verbal performances share these two "design features" with music makes a point we can all, I think, agree on. His suggestion, though, that greetings and farewells also share these two features raises even more interesting issues. If the formulaic phrases and intonation contours of greetings and farewells are modeled on a protolanguage of holistic utterances, might not this practice derive from yet older forms of social communication—namely, primate contact and location calls? After all, social networking did not originate with cell phones and iPads. Though we now attach words to our own electronic contact and location calls, their purpose is often the same—to let persons within our own "society of intimates" know where and how we are and, in return, to know where and how they are. Perhaps formulaic phrases in traditional oral narratives were similarly used to affirm cultural identity and social solidarity and did so more as instantly recognizable musical phrases than as pieces of information.

The traditional forms of oral poetry reflect more than a cultural conservatism. There is that, of course, but these paleopoetic forms seem also rooted in universal human cognitive traits. Vocal and gestural codes continue to operate alongside spoken language and continue to enrich it with fleeting nuances of mood and emotion. Moreover, the prevalence of formulaic phrasing both in spontaneous speech and in oral performance suggests that an earlier linguistic form continues to function within modern, syntax-structured language.

Epilogue
THE NEOPOETICS OF WRITING

Some 25,000 years before written language, humans were already using counting devices. These were tallies, lengths of bone or wood that were scored crosswise to record a number of items—days, months, gifts, or any other set of countable things.[1] When written language came along, it began as a means to preserve information that oral narrative and tally sticks were ill equipped to store, lists of names and places (onomasticons), census rolls, trading inventories, debt accounts, and, later, laws (Ong, 1982:99). In terms of style, lists are absolutely paratactic. That is, one item follows another just like cuts in a tally stick. It is said that oral narrative is also paratactic, one action following another, but as I pointed out earlier, this is not quite the case, for backlooping repetitions and shifts in time and place interrupt its sequential order. Nevertheless, I think we are justified in assuming that oral narratives represent a preliterate mind that, in its use of language at least, interpreted the world as a linear series of events, threadlike as the Greeks imagined each person's fated life span.

With the introduction of writing we mark the shift from the mythic to the theoretic stage, but, as we do so, we need to remind ourselves, as Merlin Donald wisely advises, that each stage, momentous as its changes were, retained all preceding stages within its modi operandi. When, therefore, we consider the impact writing had on the making of verbal artifacts, we need to discern (1) the stylistic features of oral compositions that continued unchanged, (2) their stylistic features that were modified to fit the new medium, and (3) the new stylistic features that writing set in place.

Literacy, when it first arrives in a society, is rarely shared by the entire population, but rather by a scribal elite in the service of a political elite. It is principally through oral performance arts and formal recitation that written artifacts are first distributed. Residual orality thus continues in every society to which literacy has ever been introduced. Literates, too, continue to improvise conversational discourse and store in memory songs, jokes, proverbs, and family histories, even when mnemotechnics has become a lost art. In our own time, electronic media have invigorated oral culture and produced what Walter Ong (1982) called "secondary orality," a technological extension of "primary orality," the system of preliterate information exchange.

Writing, of course, whenever and wherever it arrives, has a considerable impact on oral culture. The first and most obvious effect is the editing of the oral repertoire through the selection of particular traditional compositions for inclusion in this new storage system, leaving those not selected for transcription to fade gradually from collective memory. That was how the narratives of the Greek "epic cycle" became extinct. Prior to its transcription, an oral work would have been in a perpetual state of variation, its lines and episodes continually altered and rearranged by performers competing for audiences. The transcription of just one of its variants eventually makes it the only authoritative version. A composition that in an oral culture possesses a vital variability, its *"mouvance,"* as Paul Zumthor (1990) called it, then becomes frozen in time and space like some delicate, elaborate carving.

Though the concept of authorship is so central to literate culture, it comes down to us from an older, preliterate belief that at the dim dawn of time certain gods or divinely inspired culture heroes invented specific skills or tools. The Greeks, for example, believed that Dionysos taught humans viticulture, Triptolemos gave them the plough, Hermes the lyre, and so forth. The Romans called these the "auctores," the persons responsible for augmenting (L. *augere*) the store of human artifacts. The primary "authors" of cultural resources in a literate society also included the inventors of literary genres—e.g., Homer of epic, Hesiod of didactic verse, Thespis of drama, and Herodotus of history. Like Muses in mortal form, they represented in their paradigmatic works the ultimate authority for later writers. The ritual-derived poetic genres mentioned in chapter 7 now had fixed models for poets to imitate, ancient *auctores* that contemporary *authors* were to study night and day, as Horace advised.

When he gave that piece of advice (*De Arte Poetica*, 269–70), Horace framed it in very concrete terms: "You had better keep turning your Greek copies with nightly and daily hand" (*Vos exemplaria graeca / nocturna versate manu, versate diurna*). The *exemplaria* in question were probably copies of Greek dramas, since the addressees of Horace's epistle were the Piso brothers, aspiring playwrights, who, as upper-class Roman students, would have had their own editions of Greek works in the form of scrolls that could be unrolled manually either forward or backward. This made it possible to reuse these verbal artifacts in ways that orally situated audiences were never able to manage. Spine-bound codices, which began replacing scrolls in the first century C.E., allowed page flipping, an even more efficient way to find and reread passages.

Silent reading, while certainly possible, was not customary, for most literates found it easier to murmur the words they read. When it was convenient to do so, they read prose, as well as verse, aloud, understanding that writing, like music, was meant to be sounded. While private reading was widely practiced, oral recitation long remained the norm among literates in the Greco-Roman West. This still was a culture of public vocality, and though virtually all the texts in circulation were authored by known *scriptores*, they were written ideally to be performed orally in public. For example, when Juvenal published his first satire, he began it this way: "Am I always to be a listener (*auditor*) only, and never retaliate, I that have so often been afflicted by the Theseid of hoarse-voiced Cordus? Shall that one have recited to me (*mihi recitaverit*) his Roman comedies and this one his love lyrics—and subject me to this with impunity?"[2]

These were some of the ways oral style accommodated itself to writing or exploited the advantages writing offered. Writing, however, had unique affordances of its own that led to the crafting of genuinely *neopoetic* artifacts, properly termed "literary" works.

The invention of prose was probably the most consequential development in the history of writing, but in Greece, as Eric Havelock (1963, 1986) has shown, prose only gradually acclimated itself to the culture. Plato wrote it, but in a form modeled on the dialogues of Athenian dramatists. Herodotus' *Histories* (fifth century B.C.E.), the earliest extant work of Greek literary prose, reveals earlier oral stylistic features, or so Aristotle implies (*Rhetoric* 3.9.2) when he criticizes the historian's prose for adhering to the "continuous style" (*lexis eiromenê*). This was the storyteller's style, one event following the other in lockstep with occasional repetitions, digressions, and shifts in time and space, in short, Homeric parataxis transposed into written prose (Adrados, 2005). Writing sentences that simply added one thing after another, wrote Aristotle, was an "unpleasant" style, as tiresome as a footrace without a visible goal. He himself much preferred a periodic sentence style, one in which independent clauses were built upon dependent clauses, i.e., a hypotactic, rather than a paratactic, style.[3] Hypotaxis (literally "under-arrangement") gradually became the norm for literary and philosophic prose. It is also evident in the structure of the Aristotelian syllogism, with its first two premises, as dependent clauses, *under*lying its conclusion.

Do oral cultures think in syllogisms? The social anthropologist Jack Goody answered no: syllogistic reasoning presupposes a "graphic lay-out" (1987:221). A compound sentence, composed of equal elements strung together paratactically, cannot readily draw inferences from this series, whereas a complex sentence can successfully do so. Consider the following three examples. The first is a passage from Deborah and Barak's song celebrating the assassination of the Canaanite general, Sisera, by the Israelite woman, Jael (Hebrew Bible, Judges, 5.25–27). The second is a syllogistic rendering of the topic. The third is a rewriting of the narrative passage in the hypotactic style.

(1) Water he asked, milk she gave him;
In a lordly bowl she brought him curd.
Her hand she put to the tent-pin,
And her right hand to the workmen's hammer;
And with the hammer she smote Sisera,
she smote him through his head,
Yea, she pierced and struck through his temples.
At her feet he sunk, he fell, he lay;
At her feet he sunk, he fell;
Where he sunk, there he fell dead.[4]

(2) Major premise: All humans are mortal.
Minor premise: All Canaanites are humans.
Conclusion: All Canaanites are mortal.

(3) Women can sometimes perform great feats of courage, for example, Jael (Judges Chapter 4 & 5). Having invited the Canaanite general, Sisera, to her tent, she endeavored to keep him there, by giving him milk and a lordly bowl of curds, when he had only asked for water. When, exhausted, he fell asleep, his head resting on the table, she, having already removed a tent stake, hammered it through his temples. Though he was able to rise up briefly, he soon fell dead at her feet, having proven no match for a woman who was as clever as she was brave.

At this point I would expect someone to object to this comparison on the grounds that the three texts are totally incommensurable. Each is meant to accomplish a different end. The one is a triumphal expression of national pride extolling the courage of a particular woman. The second is an abstract generalization. The third sounds like a medieval exemplum. But that is exactly my point: oral narratives, syllogisms, and written texts are not only stylistically different, but each is constructed is such a way as to convey a different kind of information. The first, in paratactic style, represents an episode as a narrow series of actions. The second deduces an inferred conclusion from two parallel-processed premises. The third illustrates an initial generalization by referring to a biblical episode and, in hypotactic style, recounting it as a set of partially overlapping actions.

Every verbal style corresponds to a different cognitive style: when one changes, so too does the other. With that in mind, I will now conclude with some brief remarks on the cognitive evolution of the neopoetic artifact. I will begin by citing some public uses of writing and follow their changing cognitive implications as they evolve toward book literacy.

Besides list making, a privately stored form of writing, there was inscription (epigraphy), a public form used to proclaim messages or issue commands. Here words often accompanied the image of their speaker. Inscribed at the base of a statue, across a frieze, or close to a person's mouth in murals or painted ceramics, such writings functioned as disembodied voices (Svenbro, 1976, 1993). Gravestones were voices, too, that often commanded attention: "stop, wayfarer" (*siste viator*), "remember you shall die" (*memento mori*). Monuments, after all, are built to be public admonishments, the scripted voices of a past that will not stop talking (Collins, 1996:123; Gilson and Gilson, 2012).

Marshall McLuhan's (1962/2011, 1964) claim that literate cultures are visual and oral cultures auditory was at best a provocative simplification, at worst a misleading one. To clarify the cognitive implications of writing, we should first examine the cognitive processes involved in attending an oral performance. Oral performance is auditory *and* visual, listened to *and* watched (Mitchell, 2005). A rhapsode, when chanting a portion of an epic, would not only speak a third-person character's words for the audience to hear, he would also imitate him or her, through voice or visual gesture.[5] Since hearing and sight are noninterferent, these two perceptual systems can team up dyadically and process the verbal artifact concurrently, but, as is typical of such multitasking pairs, one will be more narrowly attended to, while the other receives broad, subsidiary attention. In the case of oral narration, it seems apparent that hearing is primary and vision is secondary. The visual presence and body language of the narrator will add to the illusion that the actions being described are actually taking place here and now. But, as I mentioned in the last chapter, visual perception has another cognitive consequence: it inhibits visual (mental) imaging, for while we focus on the skilled narrator's features and gestures through which he communicates indexical and iconic signs, we find it difficult to respond to his *symbolic* signs, his words, by mentally imaging them as indices and icons. This constraint on the preliterate imagination helps explain why oral epic is filled with speeches and dialogues and why it is only briefly descriptive.

Writing introduced a significantly different cognitive procedure. The written page, unlike the storyteller, is silent and motionless. The reader, unlike the audience member, is personally involved in actualizing the event. The input channel is indeed visual, but it directly leads to a decoding process that restores the graphemes to imagined speech sounds, augmented by subvocal motor innervations, the phenomenon of "inner speech." Because the written characters, unlike oral performers, are arbitrary graphic symbols signifying arbitrary linguistic symbols, they supply no indices or icons of their own. Writing therefore invites—actually, requires—the reader to imagine persons, places, and things that function as indexical and iconic signs. In short, writing exploits the brain's visual system to focus, not on the visible page, as McLuhan claimed, but on the visual images language evokes.

The development of the publishing industry in Hellenistic and Roman imperial times promoted the private consumption of written texts by (relatively) silent, solitary readers. One of the early indicators of this neopoetic trend was Aristotle's remark in his *Poetics* that tragedy can produce its intended effects even without dramatic action—simply by being read (1462a). As texts became more uniformly copied, and later, when it became customary to separate words with empty spaces, writing became more "transparent," reading more fluent, and mental imagery more feasible. The most distinctive feature of neopoetics was the power of written texts to enlarge the imaginative capacity of readers and, with that, to facilitate mental time-and-space travel. The success of scripture-based visionary religions—the "religions of the book" (Judaism, Christianity, and Islam)—was an effect, as well as a contributing cause, of the advancements in literacy from 500 B.C.E. to 700 C.E.

With the advent of the solitary, increasingly silent reader, new literary genres came into being. One was prose fiction, which mixed hyperbolic adventure with what we might now call "magical realism." This genre, which in the Late Classical West included the Greek romances (e.g., *Daphnis and Chloe*) and Latin picaresque novels (e.g., Petronius' *Satyricon* and Apuleius' *Golden Ass*), engaged readers in a new kind of verbal play, one in which the entire narrative was a fable understood to be factually untrue, yet somehow meaningful—at any rate, entertaining. The other was the lyric, the genre that by the sixteenth century C.E. would become what most readers consider the quintessential kind of poem. This latter outcome would have surprised Aristotle, who regarded tragedy as the paragon of *poiêsis* and epic a close second. He never deigned to mention the lyric, unless one interprets his passing comments on an obscure genre of Dionysian hymns, the dithyramb, as referring to lyric. This omission has provoked many centuries of critical head scratching and logical contortions, as Gérard Genette chronicled in his *Architext* (1992).

According to the standard view, the lyric originated as the text of a lyre-accompanied song. The relative brevity and strophic structure of the lyric does suggest it derives from the orally performed, musically accompanied composition. Like the song that has but one singer, the lyric is dominated by the first-person singular, addressing one or several others in *I–You* discourse and, being monologic, needs no other person's vocal input or presence. Direct address in lyric therefore often takes the form of apostrophe to an absent other or speech directed toward a personified idea or object.

The literate lyric, however, has other features that rightly place it among *literate* artifacts. It may be recited aloud, but the social setting of an oral performance is not necessary. Its venue is the mind of a solitary reader thinking the thoughts of an equally solitary writer. Both writing and reading thrive in the absence of social distractions. The typical lyric tense is the present, a fact that further strengthens the correspondence between the reader, *now* reading, and the text-represented writer, *now* uttering thoughts. Emotionally toned thoughts of the past and the future—nostalgia, regret, fears, desires—are typical of this genre, but these temporal projections are usually framed by the present. In many respects, the intimate relation of reader with writer resembles that between the addressee and the addresser of a personal letter. This early form of literate communication, as private as inscription was public, was immensely popular both in early imperial Rome and in late eighteenth-century and early nineteenth-century Europe, two periods of intense lyric production. It should also be noted that in both these periods the writers of lyric poems, even those they labeled *"carmina,"* "songs," and "odes," seldom, if ever, intended them to be set to music. (Horace's choral ode, the *Carmen Saeculare*, is a rare exception.)

The materiality of writing bestowed on the verbal artifact, be it prose or verse, a new instrumental status, affording readers new ways to pursue the old Delphic quest for self-knowledge. Now if one chose to experience mental exploration, one could reach out, grasp a book with one's hand, open it, and through the instrumentality of the written page find oneself moving along seldom-visited inner pathways. As the paths diverge and the words take one to unforeseen places, where image and motor schemas flash and vanish, one realizes that this instrument is a mental travel device.

The question that has most intrigued me about this travel is why the itinerary includes places, thoughts, feelings, and events that I have no memory of having experienced. I know, the combinatory imagination has marvelous powers, but I think there are more factors involved. This venture into paleopoetics has been based on my assumption that, despite its plasticity, the brain, like every other organ of the human body, has evolved certain structures that have enabled individuals to live long enough to pass them on to their descendants—to us who owe our existence to the fact that we are preceded by an unbroken line of sexually successful survivors. It follows that just as we have inherited from them a number of visible traits that we have never had to struggle to acquire—e.g., our opposable thumbs and binocular vision—so we have also inherited a special form of mental multitasking. Besides our ability to use figure–ground differentiation to organize a visual field of objects, we can use a similar procedure to *imagine* a scene in the absence of physical objects. Although perceiving and imagining are not identical processes, they both represent a field of objects simultaneously in broad awareness and in narrowly focused attention, a pattern I have termed "dyadic" to emphasize its integrated duality.

Why our genus is so heavily dependent on this pattern may lie in the fact that *Homo* and its one surviving subspecies, *H. sapiens sapiens*, have had to manage relatively rapid evolutionary changes. Here Dual-Process Theory, viewed from an evolutionary perspective, would suggest that System 1 represents older cognitive processes that have now come to operate peripherally within a brain newly specialized to perform centrally focused System 2 activities. The two systems, as well as each set of paired traits, I submit, replicate the dyadic pattern.

This study of the preliterate imagination, this paleopoetics, has traced a series of major reorganizations of the brain's capacity to connect

EPILOGUE: THE NEOPOETICS OF WRITING

with other brains through semiotic exchanges. These started with prehuman vocal and gestural indices, followed by iconic imitations, and then the mastery of symbolic signs vocally transmitted, semiotic adaptations that correspond to Merlin Donald's episodic, mimetic, and mythic stages. At each stage along the way, the ratchet effect (Tomasello, 2009) determined that the older mode would be retained, but modified in order to serve the newer mode. In each case, the old and the new formed a dyad in which the old mode assumed functions associated with parallel activity and broad peripheral awareness, while the new mode specialized in serial activity monitored with narrowly focused attention.

In the last two chapters I examined how language and its verbal artifacts incorporated pre-language and protolanguage as parallel-processed background features, while full, spoken language, serially produced, became the new focus of human information sharing. When we turn to consider the impact of writing on speech, we find the same dyadic pattern reemerging, but now it is oral/auditory speech that shifts to the peripheral, supporting role, while the written text becomes the dominant means of information exchange. Once again, though, the older mode has remained operative and continues to be essential to the success of the new mode of communicating.

Whether in verse or prose, verbal artifacts embody the entire sweep of evolutionary change—the process of natural selection that gave us eyes to see with and ears to hear with, cries and gestures to make known our fears and desires, and sounds to name ourselves and all the living others that we share the earth with. Reading these verbal artifacts, we need to come close enough to their words to hear the ancient pulses and tones that still resonate within them. Confronting the silent printed page, we need to imagine the sound colors of vowels and consonants, those intricate phonemes, as they first amazingly bridged the empty space between separate minds. One of the achievements of what we now call "literature" is the power it still gives us to relive that moment and, in doing so, to touch and animate that deeply living past within us.

Notes

1. The Idea of a Paleopoetics

1. The Rimbaud excerpts come from the two *lettres du voyant*: May 13, 1871, to Georges Izambard, and May 15, 1871, to Paul Demeny (Cornille and Rimbaud, 1997). The reference to T. S. Eliot comes from his 1919 essay "Tradition and the Individual Talent" (Eliot, 1957).

2. W. B. Yeats, "Sailing to Byzantium" (1927), lines 9–12.

3. Hesiod, *Theogony*, lines 1–115; Ezekiel, chaps. 1–11; Federico García Lorca, "The Theory and Function of the *Duende*" (Allen and Tallman, 1974:91–103); Graves (1948); Rilke (1922/1965).

4. The Dickinson quote is from an 1870 letter she wrote to her sometime mentor, Thomas Wentworth Higginson. Alternately bemused and astonished by her statements, the abolitionist colonel and man of letters captioned this quote as a "crowning extravaganza." It is collected in Mabel Loomis Todd, ed., *Letters of Emily Dickinson*, vol. 1 (Boston: Roberts Brothers, 1894), 315; the Ezra Pound quote is from "A Retrospect," first published in *Poetry* 1, no. 6 (March 1913), reprinted in Allen and Tallman (1974:36–54); T. S. Eliot's essay "The Metaphysical Poets," first published in 1921, is reprinted in Eliot (1957).

5. Walt Whitman, "Song of Myself," lines 110–20, 564–81, in *Leaves of Grass*, 1892 ed.

6. W. C. Williams, "Asphodel, That Greeny Flower," 1955, lines 313–26, in Williams (1966).

7. Rhetoric, as well as poetics, existed long before writing. Despite their use of notes and even teleprompters, public speakers still try to seem to be uttering their own spontaneous thoughts and feelings. Though the speeches of Demosthenes and Cicero were written in advance and have since survived to be studied as gracefully constructed works of art, they were intended to seem skillful improvisations, and to

that end they were committed to memory. Writing has allowed later generations to read and reread these monumental works of rhetorical art, but we would misread them if we did not recognize them as orally delivered reactions to unique political events.

8. Despite its title, Aristotle's *Peri Hermeneias* (*De Interpretatione*) is not about textual interpretation, but about semantics and logic. The Homeric allegorizers were principally Stoics and Neoplatonists. The commentators on Vergil were Servius, who composed line-by-line *scholia* of all the poems, and Tiberius Claudius Donatus, who wrote a general interpretation of the *Aeneid* for the benefit of his son.

9. In his comments on Peter Stockwell's *Cognitive Poetics: An Introduction*, David Miall (2005:134) notes on the very first page a "slippage in the key term 'reading,'" as "reading" (action) becomes "*a* reading" (written explication). See also Miall (2006:41–42).

10. In a follow-up article in *Poetics Today*, Richard van Oort wrote in reference to "Literature and the Cognitive Revolution": "Much of this cross-disciplinary work . . . has not produced the revolution in literary studies hoped for by its proponents. On the contrary, cognitively informed interpretations of various literary works seem for the most part content to apply the newly acquired terminology of cognitive science to the fundamentally old task of providing original interpretations of literary works. In this sense, cognitive poetics is new wine in old bottles" ("Cognitive Science and the Problem of Representation," Richard van Oort, *Poetics Today* 24, no. 2 [Summer 2003]:237–95). To be fair, I would exempt Turner and Hernadi from my own criticism, since their articles were not intent on demonstrating the interpretive uses of cognitive poetics.

11. Peter Stockwell's *Cognitive Poetics: An Introduction* (2002) exemplifies this tendency for, while it surveys current theoretical opinion with admirable thoroughness, it restricts itself to information-processing functions, introducing issues of imagination, simulation, and emotion only in its "last words" chapter.

12. His reference to "'orthodox' or 'narrow-school' EP" comes from "An Open Letter to Jonathan Kramnick," posted on the Internet in response to Kramnick's January 2011 article "Against Literary Darwinism," which appeared in *Critical Inquiry* 37, no. 2:315–47. Brian Boyd (2012) in a further attempt to distance Literary Darwinism from EP has adopted the lowercase form, "ep," to represent his brand of evolutionary psychology and has rechristened the movement "evocriticism."

13. S. M. Coleridge, *Biographia literaria*, chap. 14.

2. From Dualities to Dyads

1. Sigmund Freud also recognized this as a problem, and though he interpreted their dynamics quite differently, his "primary" and "secondary process thought"—characterizing his Pleasure Principle and his Reality Principle, respectively—closely correspond to S1 and S2. Any conflict that might arise between the two systems should not be termed a "cognitive dissonance," for, as Leon Festinger (1957) defined that principle, it would exist only between ideas or beliefs processed wholly within System 2.

2. If I could have found a term already applied to this concept, I would have used it. I imagine Arthur Koestler faced this quandary before he settled on "holon" (*Ghost in the Machine*, 1967). I had thought of "complementarity," but for various reasons, including its length, I decided against it. The term currently used, especially in computational and robotic theory, for the combination of serial and parallel processes is "hybrid," as in the "hybrid model of information processing," but, to my mind, "hybrid" implies a blending or homogenizing of the two. I prefer "dyad," because it retains the distinction between the two while integrating both in the performance of a single task. "Dyad" is often used to speak of the interaction of two persons (e.g., the "mother–child dyad") but, of course, this is not how I am using this term.

3. A sizable body of evidence has now cast doubt on the assignment of these bones to the genus *Homo*, many authorities now calling him *Australopithecus habilis*, but no one doubts that he is a close link in the human lineage.

4. To illustrate how one sport, baseball, uses throwing, clubbing, and the two grips: The thrower always uses the precision grip, while the batter always uses a two-handed power grip, except in the act of bunting, when the dominant hand typically slides upward from the neck of the bat and holds the stock with the tips of the thumb and the first two fingers. Not only does the precision grip soften the impact of the ball, it also more accurately controls the placement of the bat.

5. This grip is still technically important. Michael Patkin has attested to the skillful uses of what he called the "double grip" in making surgical incisions and suturing ("The Hand Has Two Grips: An Aspect of Surgical Dexterity," *The Lancet* [June 26, 1965] 1:1384–85). It is interesting to note that the word "surgery" derives from the Greek for "hand work," *cheirourgia*.

6. As distinct as these two modes seem in definition, this distinction has proven somewhat problematical. Treisman and Gelade (1980) claimed that visual search used both in series (first parallel, then serial). Jeremy Wolfe (1998) argued that this distinction was unnecessary, since parallel and serial operations in visual search constituted a continuum. Haslam, Porter, and Rothschild (2001) tested Wolfe's statistical experiments, found them wanting, and concluded that the two processes were indeed distinct and that, if a continuum could be found, it would have to accommodate that distinction.

7. "Enactive perception," which alludes to a controversial idea that derives from James Gibson's "ecological optics" (1979), was reintroduced by Alva Noë (2004) and critiqued by Jesse Prinz (2006).

8. Cf. William Wordsworth's "wise passiveness" in "Expostulation and Reply," line 24, and Virginia Woolf's short story "Kew Gardens," which Dainton's meditation closely resembles.

9. Mammalian emotion has evolved as an "increasingly flexible adaptation to environmental contingencies by decoupling stimulus from response and thus creating a latency time for response optimization" (Scherer, 2001:92). Though Scherer stresses the sequentiality of this "checking" process, he acknowledges that, being "multilevel," emotions may also be considered parallel processes.

NOTES

3. Play and Instrumentality

1. "This," he adds, "is the basis of Baldwin's effect." This effect, named for the philosopher and evolutionist James Mark Baldwin (1861–1934), is an adjustment of behavior that allows some individuals and communities to survive in the face of novel circumstances. It is a deliberate adaptation that preserves genetic variations in a given population.

2. With language emerges what Donald calls the "hybrid mind," which is partly analog and partly symbolic (2001:155). The adjective "hybrid" here corresponds to what I have called "dyadic." As for its genetic origin, Donald argues against the modular theories associated with Chomsky and Fodor. Calling his view "biocultural," he places the "origin of language in cognitive communities, in the interconnected and distributed activity of many brains. . . ." (2001:252).

3. In his response to the critics who say that, unless a child is old enough to explain in words that her behavior is pretense, we cannot assume it is, Leslie (1987) argues that if her behavior demonstrates that one object has been made to stand for another, "we have reason to believe that the child is pretending" (414). I am applying this commonsensical criterion to cats, dogs, otters, crows, etc.

4. Developmental psychologists, following Piaget's lead, generally refer to the use of objects in pretend play appearing during the period 18–24 months as "symbolic play." Since semiotic distinctions are critically important to my analysis, I find this use of "symbol" insufficiently precise since all it means is "sign." When, later on, I speak of language in terms of *symbolic signs* and reintroduce the play principle, I will be concerned with the way arbitrary signs—true semiotic symbols—become the elements of human play. Accordingly, children's play objects at this stage are almost always icons, e.g., banana for telephone.

5. According to the semiotics of Charles S. Peirce, a thing functions as an *index* when a perceiver interprets it in relation to a real-world context. For example, if this thing is part of X or is a cause or effect of X, it can be used indexically to signify X. A thing functions as an *icon* when the perceiver regards it as a representation of X based on similarity and not on any contextual connection. A thing—e.g., a spoken or written word—functions as a *symbol* if its representation of X is based solely on mutual agreement or social convention.

6. For discussions of the mirror neuron system and its implications, see Stamenev and Gallese (2002), especially the articles by Fogassi and Gallese, Rizzolatti et al., Voegeley and Newen, Li and Hombert, Studdert-Kennedy, Stamenev, Bichakjian, and Morrison. Another valuable collection is Hurley and Chater (2005a), especially the separate articles by Gallese and Hurley. I will further discuss mirror neurons in later chapters.

7. To be precise, Donald (2001:260) estimates that the mimetic peaked in *H. erectus* 2–0.4 mya and was followed thereafter by the mythic stage of *H. sapiens sapiens* 0.5 mya–present.

8. This is not to say that Lower and Middle Paleolithic stone technology showed much innovation over time. It was remarkable less for its inventiveness than for its product standardization. Perhaps these early humans expressed their inventiveness in other materials or behaviors, but not in the style of their stone tools.

9. In regard to this object–instrument distinction, Napier would appear to agree that, when an object is grasped purposively and becomes a tool, it becomes an extension of the user's central nervous system, which is firmly prewired for these, and *only* these two, grips (1980:905, 913).

10. Victor Egger, *La Parole Intérieure*: "... dans tout jeu, dans toute feinte, l'âme se dédouble, et l'acteur convaincu recouvre un spectateur sceptique.... [D]ans le jeu, d'une façon générale, le moi individuel s'affirme et se nie simultanément ou à des intervalles indiscernables.... Ce faisant, l'esprit ne croit pas se contredire: de cette affirmation et de cette négation il fait le synthèse, et de cette synthèse est l'idée même du jeu et du drame."

4. The World as We See It

1. The numbers of degrees that appear in the literature of visual perception vary. The numbers I use are median estimates and should be considered approximate. Fortunately for my purposes, the variation is not a significant factor.

2. Michael Posner (1980) used a variation of it when he spoke of visual attention as a movable spotlight.

3. Thoreau, *Journals*, Sept. 13, 1852; June 14, 1854. For further "side of the eye" references, see also his entries for April 28 and 30, 1856 (Thoreau, 1962). Hermann von Helmholtz (*Handbook of Physiological Optics*, 1866) maintained that attention can be voluntarily shifted within the peripheral field, i.e., without directing focal vision onto an object. William James endorsed this view in *The Principles of Psychology*, vol. 1 (James, 1890/1950:435–39). See also Aristotle, *Meteorologica*, 1.6.

4. Jakob von Uexküll, *Umwelt und Innenwelt der Thiere* (Berlin: Springer Verlag, 1921).

5. Needless to say, I am grossly simplifying the intricacy of the circuitry, e.g., the manner in which the information from the left and right hemisphere of each retina is distributed to both cerebral hemispheres, the function of the two lateral geniculate nuclei, and the processes performed by the various areas anterior to the primary visual center, V_1.

6. This process of selection and recognition leads us to consider several further implications. The first is that, if selection must precede recognition sequentially, the processing that, up to this point, has been parallel now seems serial. This switchover suggests that knowing what is out there in the world can often require the sort of effort that only serial processing, albeit swiftly executed, can accomplish. For insight into the complexity of this issue, see Jean Bullier and Lionel G. Nowak (1995). Two other points are also worth considering. One is that Jeannerod and Jacob's term "selection" introduces into neuroscience the concept of "figure and ground," central to Gestalt psychology and phenomenology. The other is that "recognition" (or "object recognition") reintroduces the Kantian concept of "apperception," a principle of psychology that was generally uncontested from the mid-nineteenth century to the first two decades of the twentieth century. At least some of the "surmises" of philosophers and introspectionists, long scoffed at by the

NOTES

behaviorists and their allies, have now, it seems, been renamed and rehabilitated by a new generation of empiricists.

7. They determined the cognitive function of each stream by analyzing the visual capacity of patients who had lost the use of the *other* stream and then by applying the principle of "double dissociation." That is, patients suffering from agnosia, like D. F., whose ventral stream was severely impaired, revealed the functions of the now neurally dissociated *dorsal* stream. Likewise, patients with an injured dorsal stream and suffering from optic ataxia revealed the inherent capacities of the *ventral* stream. In neither of such instances was there the possibility of crosstalk between streams. See Milner and Goodale, 1995:92–101, 120–47.

8. The literature on frames of reference includes a variety of synonymous terms: the *allocentric* frame has been called "object-to-object," "world-centered," "environment-centered," "geocentric," "intrinsic," and "categorical"; the *egocentric* has been called "self-to-object," "body-centered," and "coordinate."

9. The assignment of mental (or cognitive) mapping to the spatial frames has been somewhat controversial. Milner and Goodale (1995) flatly state that only allocentric coding could maintain a mental map (90–91). In a well received article Wang and Spelke (2002) held that viewpoint-dependent, spatially updated egocentric coding is all that is needed. Since the two streams, each with its own frame, must of necessity work simultaneously, I find Neil Burgess's (2006) complementary model more convincing: "egocentric representations exist in parallel to (rather than instead of) allocentric ones."

10. "Massive" is quite the attention-grabber, applied as it often is to catastrophic events such as earthquakes, firestorms, and heart attacks. Daniel Dennett was probably most responsible for popularizing it when he called the brain a "massively parallel processing machine" (*Consciousness Explained* [London: Penguin Books, 1992], 127). He was not the first to use it, though. The earliest usage I found dated from 1981, when it was applied to new computer technology.

11. "Pathway" was the preferred image for Ungerleider and Mishkin and for those who continued to speak of the dorsal as representing spatial perception (the "where?" pathway). "Stream," which is now the more widely accepted usage, is associated with researchers such as Milner, Goodale, and Jeannerod. How ongoing research into the non-neuronal glial cells may affect the way we think of neurotransmission is not yet clear.

12. These transitive parts of consciousness are analogous to what Bohr called "quantum leaps," events when an electron transits from one orbit around the nucleus of an atom, to another orbit. Like them also, these transitions in consciousness, according to James, cannot be observed without interfering with them. To try to do so is like trying to hold a snowflake, "seizing a spinning top to catch its motion, or trying to turn up the gas [light] quickly enough to see how the darkness looks" (James, 1890/1950:245), observations that anticipate the Uncertainty Principle later formulated by Bohr's student, Werner Heisenberg.

5. Human Communication: From Pre-Language to Protolanguage

1. The main thesis of Dunbar's 1996 book, *Grooming, Gossip and the Evolution of Language* (London: Faber & Faber), was that, as hominids came to live in communities far larger than their ape ancestors, the social obligation to groom all those they needed to bond with individually became too time-consuming, so the vocal soothing of several at a time took its place. Citing sociological data from a number of different cultures, Dunbar found that the average length of time that humans daily engage in gossip today roughly correlates to the time that great apes spend per day in grooming one another's fur.

2. For a much more nuanced evaluation of the social function of ecological (object) information, see Kim Sterelny, "Social Intelligence, Human Intelligence and Niche Construction," *Philosophical Transactions of the Royal Society B* (2007):719–30. For an early assessment of Bickerton's stance on language origins, see a review of his *Language and Species* by Michael Studdert-Kennedy (1991), who critiques Bickerton's embrace of the "catastrophic theory" and his view of communication: "[I]nstead of treating the communicative and representational functions as mutually reinforcing components of a feedback system—the more you say, the more you have to say, and vice versa—he disregards the communicative function almost entirely" (259, 261).

3. Such was the conclusion that Philip Lieberman arrived at as early as 1971. Lieberman has always found improbable the notion of the sudden genetic mutation, a saltation, in which *Homo sapiens sapiens* acquired what Noam Chomsky called the human "language organ" and Stephen Pinker the "language instinct." The Neanderthals, living cooperatively in small bands, probably had language of some sort, just not the sort that our direct ancestors had (Philip Lieberman, *The Biology and Evolution of Language* [Cambridge, Mass.: Harvard University Press], 1984). The relation between the two species continues to be redefined: recent DNA analysis has found evidence that Neanderthals and modern humans may have interbred as early as 80,000 years ago when their populations met in the Middle East. If Caucasians prove to be genetically related to *Homo neanderthalensis*, this species did not become extinct, but rather merged with *Homo sapiens sapiens*, a fact that, no doubt, will have profound implications for the study of biological and cultural evolution. See Richard E. Green et al., "A Draft Sequence of the Neandertal Genome," *Science* 328, no. 5979 (May 7, 2010):710–22.

4. On abbreviation as a factor in the evolution of sign systems, see Brian MacWhinney, "The Gradual Emergence of Language," 245 (in Givón and Malle, 2004); Robbins Burling, "Comprehension, Production, and Conventionalization in the Origins of Language," 27–39 (in Knight et al., 2000); and Michael C. Corballis, "Did Language Evolve from Gestures?" 163–64 (in Wray, 2002b).

5. Michael Arbib (2009a) has accepted Wray's theory as currently the most plausible model for a protolanguage, or as he calls it, "protospeech" (to balance the concept of "protosign," i.e., gesture). Cf. Bronaslaw Malinowski's notion of "phatic utterance."

6. Common English usage preserves a special connection between signs and visual cognition. We "see" a sign and absorb the information it sets forth. Unless

NOTES

the context of our conversation happens to be semiotics, we never say we "hear" a sign. Instead, we "hear" a *signal*. Signals may be auditory or visual, but whatever their modality they are not messages to be pondered or perused: they carry simple meanings, often cues to immediate action. See Winfried Nöth (1995:111–12).

7. "Admittedly," Deacon acknowledges, "this is not the way we typically use the term iconic, but I think it illuminates the most basic sense of the concept" (1997:75). In his review of the book, Richard Hudson (*Journal of Pragmatics* 33 [201]:129–35) remarks that "he seems to me to use 'icon' where other people simply talk of categorization" (130).

8. Chimps and bonobos can learn to point, once they have been enculturated by human caregivers, but they do not do so in the wild. It is interesting to note that these manually adept apes *can* make these indexical gestures but apparently have no need to do so in their natural habitats.

9. After discussing the role of "linguistic indexicals (e.g., *that*) and shifters (e.g., *you*)" as symbolic signs that point, Deacon (2003:134) comments: "It should not go unnoticed that this is consistent with arguments suggesting an evolutionary development of spoken language from ancestral forms that were entirely or partially manual." Among those who should notice this diplomatically worded aside would presumably be Michael Arbib, Michael Corballis, and Robin Dunbar, whose articles appear in the same collection.

10. As a vocal medium, language exhibits both the weaknesses and strengths of a "short form." Its phonemes and the words they construct are conventional features that rarely sound anything like their meanings (Hockett, 1960/1982). Speech is therefore inherently unclear, unless it is grounded in perceptual and conceptual contexts, i.e., the immediate setting of the speech event and the preestablished topic of the ongoing discourse. Once grounded, though, words prove to be extremely flexible means of communication. Their flexible ambiguity is the result of a "trade-off between two communicative pressures which are *inherent to any* communicative system: *clarity* and *ease* [cf. my "long" and "short form"]. A *clear* communication system is one in which the intended meaning can be recovered from the signal with high probability. An *easy* communication system is one in which signals are efficiently produced, communicated, and processed" (Piantadosi et al., 2012:281, authors' emphasis). Wray and Grace's (2007) protolanguage would accordingly be classified as an "easy," or short form, system—esoteric, holistic, and formulaic. Their full language would be a "clear," or long form, system—exoteric, syntactical, and compositional.

11. In brief, "exaptation" is the process by which a trait originally shaped by natural selection to serve one purpose is co-opted for a wholly new use. See Stephen J. Gould and Elizabeth Vrba, "Exaptation—A Missing Term in the Science of Form" (Gould and Vrba, 1982).

12. For the possible relation of the breathing and chewing cycles to the evolution of language, see Peter F. MacNeilage, "Whatever Happened to Articulate Speech?" (in Corballis and Lee, 1999:116–37).

13. The elbow, wrist, and opened hand motion he describes resembles a miniature act of throwing. On throwing as an evolutionary factor associated with pointing, see William Calvin (1993, 2004).

14. Our tendency to use visual clues (lip reading) to help process vocal speech has been explored by Harry McGurk and J. McDonald (1976). This phenomenon (the "McGurk Effect") continues to intrigue speech scientists.

15. It is worth noting that Jean-Jacques Rousseau in his *Discourse on the Origin of Inequality Among Men* (1754) anticipated Wray's theory and my addendum: "It is reasonable to suppose that the words first made use of by mankind had a much more extensive signification than those used in languages already formed, and that ignorant as they were of the division of discourse into its constituent parts, they at first gave *every single word the sense of a whole proposition*. When they began to distinguish subject and attribute, and noun and verb, which was itself no common effort of genius, *substantives were first only so many proper names*" (Rousseau, 1754/1984, part 1, my emphasis).

6. Language: Its Prelinguistic Inheritance

1. Cf. André Martinet's (1964) concept of "double articulation." Hockett's other three were displacement (the ability to refer to absent or imaginary objects), productivity (the ability to compose novel, yet comprehensible, statements), and traditional transmission (the ability to learn the special features of one's native language). Hockett was careful not to make absolute claims as to the uniquely human character of two of these final four. For example, displacement is "apparently almost unique"—the bees do dances about absent flowers, though chimps do not communicate notions of absent entities (Hockett, 1960/1982:6). As for traditional transmission, he leaves open the possibility that gibbons' calls may be "extragenetically" learned.

2. This form of play uses symbolic *signs* and needs therefore to be distinguished from what developmental psychologists refer to as "symbolic play," which from a semiotic point of view might best be called either *indexical* or *iconic* play. When developmental psychologists speak of "symbolic play," they refer to objects that children have imposed their own meanings on—the blanket that is a mother substitute (Winnicott) or the banana that is treated as a telephone (Leslie). As I've pointed out earlier, such objects are not semiotic symbols, but connote what Freud and Jung meant by the word, namely, subjectively charged (sometimes "cathected" and "overdetermined") objects or images.

3. As the writings of George Lakoff, Mark Johnson, and Mark Turner have revealed, the vocabulary of modern languages, and presumably of early, no-longer-available languages as well, are substantially formed through metonymic and metaphoric processes. That is to say, the meanings of most nouns and verbs derive from other words to which they are associated by *contiguity* (cause/effect or part/whole relation) or by *similarity* (shared properties or isomorphy).

4. I refer here to concrete, common, count nouns, which I take to be the earliest and still the basic-level designations for objects. I do not refer to mass nouns or abstract concepts, which I assume to be derivatives of count nouns.

5. I refer here, of course, to that class of prepositions that spatially locate objects relative to other objects, not to such conceptual prepositions as *without, according to, because of,* and *except*, etc., or to prepositions used as verb complements

and infinitives. Saccadic suppression (or masking) was first reported by Raymond Dodge in 1900. For a somewhat fuller discussion of prepositions and imaging, see Collins, 1991a:115–18.

6. Primate visual perception is many millions of years older than language, but did language when it came along affect visual perception? Moreover, do separate languages determine how their speakers process visual information? (This is, of course, the linguistic relativism of the Sapir-Whorf Hypothesis.) Or does the visual system that all humans have inherited explain why separate languages are similar? Though this debate continues, it seems that a reasonable answer to each debating team is a qualified "yes." As Regier and Kay (2009) would have it, both are "half right."

7. When Langacker and other cognitive linguists refer to visual perception, they often simply use the word "perception," an understandable abbreviation, but one that implicitly excludes other sense modalities and tends to minimize phonology and audition as linguistic factors.

8. See Michael Arbib (2008) for a more extended discussion of this issue and a sharply reasoned critique of Gallese and Lakoff (2005).

9. Talmy is aware of Ungerleider and Mishkin's (1982) findings on the two visual pathways, acknowledging that the "what" and "where" systems fit his two subsystems quite well, but he feels that *structure*, while broadly locational, relies on the structural representation of single objects, not just an "extended object array" (Talmy, 2000:167–68).

10. The authors of the last cited paper, an empirical study of spatial semantics, seem to have taken it for granted that Talmy could not have meant what he said about "observer-neutral" emanations, for they misinterpret his assessment of fictive motion as originating "in the 'perceiver' of the event who mentally 'scans' or 'goes through' that mental space" (Wallentin et al., 2005:222).

11. "From at least the time of Quintilian, 'common places' meant both the places the arguments were stored in and the arguments themselves" (Ong, 1982:110–11). In referring to topics as arguments stored in *topoi*, I follow that tradition. "Commonplace," the adjective, has come to mean "trite," because rhetorical overuse made the practice all too predictable.

7. The Poetics of the Verbal Artifact

1. "Semiotic constraints delimit the outside limits of the space of possibilities in which languages have evolved within our species, because they are the outside limits of the evolution of any symbolic form of communication. So perhaps the most astonishing implication of this hypothesis is that we should expect that many of the core universals expressed in human languages will of necessity be embodied in any symbolic communication system, even one used by an alien race on some distant planet!" (Deacon, 2003:138).

2. This ritual hypothesis for the social origins of the verbal artifact contrasts with the sexual selection hypothesis advanced in one form or another by Literary Darwinists. The latter argue that verse, as distinct from prose narrative, is a male

strategem used to attract females who interpret it as an indicator of verbal cleverness. Like the peacock's tail, a cumbersome extravagance that charms peahens while rendering the peacock vulnerable to predators, poetry, with its "extremely high tolerance for expressive nonsense" (Vermeule, 2012:429), is "merely what Daniel Dennett calls a 'good trick'" (Boyd, 2012:401).

3. Robert Lowth (1710–1787) is credited with having first identified biblical parallelism as a poetic feature. Subsequent research has discovered that it is also a feature of Ugaritic poetry, composed in an older Semitic language related to Canaanite and Phoenician (Berlin, 1992).

4. In his article, Russo distinguished each type with a different form of underscoring. To simplify matters, I italicized his marked repetitions and used virgules to distinguish separate kinds of repetition. For his more particular textual analysis, I very much recommend the entire article.

5. Paivio's theory is no longer considered current, but it was, early on, important to the research of Stephen Kosslyn and his colleagues. Paivio's early opposition to the then influential modular theory of the mind and to the notion of a computational "mentalese" has proved justified in light of recent work on conceptualization and visuomotor simulation (Barsalou, 2009).

6. Tulving (2002), citing functional brain imaging research published in the 1990s, reported that during episodic encoding the left prefrontal cortex is principally involved, whereas during retrieval the right prefrontal cortex is activated. This suggested to him that the latter is involved in time travel back to the time and place of the encoding (17–18).

7. David Burrows has commented on the tendency, in song, of the music overriding the words. Examples of this include the "prolongation of vowel sounds, repetitions of certain words and phrases, introduction of rests all [of which] may stretch the normal time span for taking in sentences past the breaking point" (1990:88). In some performance traditions, as in post-Renaissance Western singing, this may be characterized as music cannibalizing its verbal partner, a case of "logophagia" (ibid.:87). In other traditions—e.g., plain song and Quranic recitation—the words and their phrasal structures are carefully articulated, with melody used only to heighten their effect. Related *parlando* styles are operatic *recitativo*, *Sprechgesang*, and *Sprechstimme*. In orally performed narrative, the voice carefully articulates the verbal message and if music, vocal and/or instrumental, accompanies the words, it functions as ground, not as figure.

8. See my *Reading the Written Image* (1991b:18–21) and *Poetics of the Mind's Eye* (1991a:2–9).

9. This connection does not seem to have attracted much attention from classical scholars. Wray herself only mentions in passing the Parry-Lord theory (Wray, 2002a:75–76). But one young scholar, Chiara Bozzone (2010), taking a cognitive approach, has developed connections that other classicists should find well worth considering.

NOTES

Epilogue: The Neopoetics of Writing

1. A (pre)history of tallies would, if written, have much to tell us of the origins of storytelling, I suspect, but for now it is enough to observe that many of our words for writing derive from words for making scratches, e.g., Gr. *graphein*, L. *scribere*, O.E. *writan*, O.H.G. *rizan*. This may simply mean that speech was first transcribed by scratching letters onto a surface, or it may mean that, once a pictographic, syllabic, or alphabetic system was invented, it seemed reasonable to consider it an extension of that long-established method of notation, tallying. Newfangled voice-transcription retained the old name, "scratching," perhaps for the same reason that digital computing skill uses names such as "desktop," "folders," and "trash."

2. *Semper ego auditor tantum? numquamne reponam /vexatus totiens rauci Theseide Cordi? / inpune ergo mihi recitaverit ille togatas, / hic elegos?* Lines like these remind one nowadays of a stand-up comic's routine—a very clever comic with attitude who writes his own material. The *Theseid* is apparently an epic poem featuring Theseus as hero. Note: we still refer to what a writer is "saying" and a writer's readership as his or her "audience."

3. "Period," from *peri* + *hodos*, a pathway around, a circuit. For a discussion of a related stylistic issue raised by Aristotle in his Rhetoric, "written style" (*lexis graphikê*) as distinct from other rhetorical styles, see Graff, 2001. See also Collins, 1991a:68–76, 83–84, and 102.

4. My source is the 1917 translation provided by the Jewish Publication Society.

5. Plato, *Republic* 353c (*kata phônên hê kata schêma*). Epic was therefore classified as a mix of diegesis and mimesis—the "mixed mode." Since subvocal mirroring of heard words facilitates a hearer's understanding of them, we should add motor response to this mix.

Bibliography

Abry, C., A. Vilain, and J.-L. Schwartz, eds. 2009. *Vocalize to Localize*. Amsterdam: John Benjamins.
Adrados, F. R. 2005. *A History of the Greek Language: From Its Origins to the Present*. Leiden, Netherlands: Brill.
Aiello, L., and R. Dunbar. 1993. "Neocortex Size, Group Size and the Evolution of Language." *Current Anthropology* 34:184–93.
Allen, D., and W. Tallman. 1974. *Poetics of the New American Poetry*. New York: Grove Press.
Arbib, M. A. 2008. "From Grasp to Language: Embodied Concepts and the Challenge of Abstraction." *Journal of Physiology—Paris* 102:4–20.
Arbib, M. A. 2009a. "Evolving the Language-Ready Brain and the Social Mechanisms that Support Language." *Journal of Communication Disorders* 42:263–71.
Arbib, M. A. 2009b. "Interweaving Protosign and Protospeech: Further Developments Beyond the Mirror." In *Vocalize to Localize*, edited by C. Abry, A. Vilain, and J.-L. Schwartz, 107–32. Amsterdam: John Benjamins.
Arbib, M. A. 2010. "Mirror System Activity for Action and Language Is Embedded in the Integration of Dorsal and Ventral Pathways." *Brain & Language* 112: 12–24.
Arbib, M. A., and D. Bickerton, eds. 2010. *The Emergence of Protolanguage: Holophrasis vs. Compositionality*. Amsterdam: John Benjamins.
Armstrong, D. F., W. C. Stokoe, and S. E. Wilcox. 1995. *Gesture and the Nature of Language*. New York: Cambridge University Press.
Arnheim, R. 1961. "Perceptual Analysis of a Cosmological Symbol." *The Journal of Aesthetics and Art Criticism* 19:389–99.
Barkow, J., L. Cosmides, and J. Tooby, eds. 1992. *The Adapted Mind: Evolutionary Psychology and the Generation of Culture*. New York: Oxford University Press.

Barsalou, L. W. 2008. "Grounded Cognition." *Annual Review of Psychology* 59:617–45.
Barsalou, L. W. 2009. "Simulation, Situated Conceptualization, and Prediction." *Philosophical Transactions of the Royal Society B* 364:1281–89.
Bateson, G. 1972. "A Theory of Play and Fantasy." In his *Steps to an Ecology of Mind*, 177–93. New York: Chandler.
Bauer, D. M., and S. J. McKinstry. 1991. *Feminism, Bakhtin, and the Dialogic*. Albany: State University of New York Press.
Bax, M. 2009. "Generic Evolution: Ritual, Rhetoric, and the Rise of Discursive Rationality." *Journal of Pragmatics* 41:780–805.
Bekoff, M., and J. A. Byers. 1998. *Animal Play: Evolutionary, Comparative, and Ecological Perspectives*. Cambridge, UK: Cambridge University Press.
Belin, P., and R. J. Zatorre. 2000. "'What,' 'Where,' and 'How' in Auditory Cortex." *Nature Neuroscience* 3:965–66.
Bell, D. V. J. 1975. *Power, Influence, and Authority: An Essay in Political Linguistics*. New York: Oxford University Press.
Benveniste, E. 1966/1973. *Problems in General Linguistics*. Coral Gables, Fla.: University of Miami Press.
Berlin, A. 1992. *The Dynamics of Biblical Parallelism*. Bloomington: Indiana University Press.
Bickerton, D. 1992. *Language & Species*. Chicago: University of Chicago Press.
Bickerton, D. 2002. "Foraging Versus Social Intelligence in the Evolution of Protolanguage." In *Transition to Language: Studies in the Evolution of Language*, edited by A. Wray, 207–25. New York: Oxford University Press.
Bickerton, D. 2007. "Language Evolution: A Brief Guide for Linguists." *Lingua* 117:510–26.
Blumenthal, A. L., ed. and trans. 1970. *Language and Psychology: Historical Aspects of Psycholinguistics*. New York: Wiley.
Bower, G. H. 1970. "Analysis of a Mnemonic Device: Modern Psychology Uncovers the Powerful Components of an Ancient System for Improving Memory." *American Scientist* 58(5):496–510.
Boyd, B. 2009. *On the Origin of Stories: Evolution, Cognition, and Fiction*. Cambridge, Mass.: Belknap Press of Harvard University Press.
Boyd, B. 2012. "For Evocriticism: Minds Shaped to Be Re-Shaped." *Critical Inquiry* 38(2):394–404.
Boyer, P., and C. Ramble. 2001. "Cognitive Templates for Religious Concepts: Cross-Cultural Evidence for Recall of Counter-Intuitive Representations." *Cognitive Science* 25:535–64.
Bozzone, C. 2010. "New Perspectives on Formularity." In *Proceedings of the 21st Annual UCLA Indo-European Conference*, edited by S. W. Jamison, H. C. Melchert, and B. Vine, 27–44. Bremen, Germany: Hempen.
Brandt, L. 2009. "Subjectivity in the Act of Representing: The Case for Subjective Motion and Change." *Phenomenology and the Cognitive Sciences* 8(4):573–601. DOI:10.1007/s11097-009-9123-9.

Braun, J., and D. Sagi. 1990. "Vision Outside the Focus of Attention." *Perception & Psychophysics* 48:45–58.

Broadbent, D. 1958. *Perception and Communication*. Oxford: Pergamon Press.

Bruner, J. S. 1972. "The Nature and Uses of Immaturity." *American Psychologist* 27:687–708.

Buller, D. J. 2006. *Adapting Minds: Evolutionary Psychology and the Persistent Quest for Human Nature*. Cambridge, Mass.: MIT Press.

Bullier, J. 2003. "Hierarchies of Cortical Areas." In *The Primate Visual System*, edited by J. H. Kaas and C. E. Collins, 181–204. London: Taylor & Francis (CRC Press).

Bullier, J., and L. G. Nowak. 1995. "Parallel Versus Serial Processing: New Vistas on the Distributed Organization of the Visual System." *Current Opinion in Neurobiology* 5:495–503.

Burgess, N. 2006. "Spatial Memory: How Egocentric and Allocentric Combine." *Trends in Cognitive Sciences* 10:551–57.

Burke, K. 1973. *The Philosophy of Literary Form: Studies in Symbolic Action*. Berkeley: University of California Press.

Burkert, W. 1983. *Homo Necans: The Anthropology of Ancient Greek Sacrificial Ritual and Myth*. Berkeley: University of California Press.

Burkert, W., and F. Graf, eds. 1998. *Ansichten griechischer Rituale: Geburtstags-Symposium für Walter Burkert, Castelen bei Basel, 15. bis 18. März 1996*. Stuttgart: B.G. Teubner.

Burling, R. 2000. "Comprehension, Production, and Conventionalization in the Origins of Language." In *The Evolutionary Emergence of Language Social Function and the Origins of Linguistic Form*, edited by C. Knight, M. Studdert-Kennedy, and J. R. Hurford, 27–39. Cambridge, UK: Cambridge University Press.

Burrows, D. L. 1990. *Sound, Speech, and Music*. Amherst: University of Massachusetts Press.

Buss, D. M., ed. 2005. *The Handbook of Evolutionary Psychology*. Hoboken, N.J.: Wiley.

Byrne, P., S. Becker, and N. Burgess. 2007. "Remembering the Past and Imagining the Future: A Neural Model of Spatial Memory and Imagery." *The Psychological Review* 114:340–75.

Byrne, R. 1999. "Human Cognitive Evolution." In *The Descent of Mind: Psychological Perspectives on Hominid Evolution*, edited by M. C. Corballis and S. E. G. Lea, 80–81. New York: Oxford University Press.

Byrne, R. W., and A. Whiten, eds. 1988. *Machiavellian Intelligence: Social Expertise and the Evolution of Intellect in Monkeys, Apes, and Humans*. New York: Oxford University Press.

Call, J., and M. Tomasello. 2006. "Apes: Gesture Communication." In *Encyclopedia of Language & Linguistics*, edited by K. Brown, 317–21. Amsterdam: Elsevier.

Calvin, W. 1993. "The Unitary Hypothesis: A Common Neural Circuitry for Novel Manipulations, Language, Plan-Ahead, and Throwing." In *Tools, Language, and*

BIBLIOGRAPHY

Cognition in Human Evolution, edited by K. R. Gibson and T. Ingold, 230–50. Cambridge, UK: Cambridge University Press.

Calvin, W. 2004. *A Brief History of the Mind*. New York: Oxford University Press.

Carroll, J. 1995. *Evolution and Literary Theory*. Columbia: University of Missouri Press.

Carroll, J. 1999. "Wilson's *Consilience* and Literary Study." *Philosophy and Literature* 23(2):393–413.

Carroll, J. 2004. *Literary Darwinism: Evolution, Human Nature, and Literature*. New York: Routledge.

Carroll, J. 2005. "Evolutionary Psychology and Literary Study." In *The Handbook of Evolutionary Psychology*, edited by D. Buss, 931–52. Hoboken, N.J.: Wiley.

Cave, K. R., and J. M. Wolfe. 1990. "Modeling the Role of Parallel Processing in Visual Search." *Cognitive Psychology* 22:225–71.

Chater, N., and M. H. Christiansen. 2010. "Language Acquisition Meets Language Evolution." *Cognitive Science* 34:1131–57.

Child, F. J., H. C. Sargent, and G. L. Kittridge. 1904. *English and Scottish Popular Ballads: Edited from the Collection of Francis James Child*. Boston: Houghton, Mifflin.

Chomsky, N. 1968. *Language and Mind*. New York: Harcourt, Brace & World.

Chomsky, N. 1980. *Rules and Representations*. Oxford: Basil Blackwell.

Christian, D. 1991. "The Case for 'Big History.'" *Journal of World History* 2:223–38.

Christiansen, M. H., and N. Chater. 2008. "Language as Shaped by the Brain." *Behavioral and Brain Sciences* 31:489–558. DOI:10.1017/S0140525X08004998.

Christiansen, M. H., and S. Kirby, eds. 2003. *Language Evolution*. New York: Oxford University Press.

Clark, A. 2009. "Perception, Action, and Experience: Unraveling the Golden Braid." *Neuropsychologia* 47(6):1460–68. DOI:10.1016/j.neuropsychologia.2008.10.020.

Cohen, B. H. 1986. "The Motor Theory of Voluntary Thinking." In *Consciousness and Self-Regulation*, vol. 4, edited by G. E. Schwartz, R. J. Davidson, and D. Shapiro, 19–54. New York: Plenum Press.

Collins, A. M., and E. F. Loftus. 1975. "A Spreading Activation Theory of Semantic Processing." *Psychological Review* 82:407–28.

Collins, A. M., and M. R. Quillian. 1969. "Retrieval Time from Semantic Memory." *Journal of Verbal Learning and Verbal Behavior* 8:240–47.

Collins, C. 1991a. *Poetics of the Mind's Eye: Literature and the Psychology of Imagination*. Philadelphia: University of Pennsylvania Press.

Collins, C. 1991b. *Reading the Written Image: Verbal Play, Interpretation, and the Roots of Iconophobia*. University Park, Pa.: Penn State University Press.

Collins, C. 1996. *Authority Figures: Metaphors of Mastery from the Iliad to the Apocalypse*. London: Rowman & Littlefield.

Confer, J. C., J. A. Easton, D. S. Fleischman, C. D. Goetz, D. M. G. Lewis, C. Perilloux, and D. M. Buss. 2010. "Evolutionary Psychology: Controversies, Questions, Prospects, and Limitations." *American Psychological Association* 65(2):110–26. DOI:10.1037/a0018413.

Corballis, M. C. 1993. *The Lopsided Ape: The Evolution of the Generative Mind*. New York: Oxford University Press.

Corballis, M. C. 2002. *From Hand to Mouth: The Origins of Language*. Princeton, N.J.: Princeton University Press.

Corballis, M. C. 2011. *The Recursive Mind: The Origins of Human Language, Thought, and Civilization*. Princeton, N.J.: Princeton University Press.

Corballis, M. C., and S. E. G. Lea, eds. 1999. *The Descent of Mind: Psychological Perspectives on Hominid Evolution*. New York: Oxford University Press.

Cornille, J.-L., and A. Rimbaud. 1997. *L'Épître du voyant: Alcide Bava/Arthur Rimbaud*. Amsterdam: Rodopi.

Coulson, S. 2001. *Semantic Leaps: Frame-Shifting and Conceptual Blending in Meaning Construction*. Cambridge, UK: Cambridge University Press.

Coulson, S., and T. Oakley. 2005. "Blending and Coded Meaning: Literal and Figurative Meaning in Cognitive Semantics." *Journal of Pragmatics* 37:1510–36.

Dainton, B. 2000. *Stream of Consciousness: Unity and Continuity in Conscious Experience*. New York: Routledge.

Darlow, A. L., and S. A. Sloman. 2010. "Two Systems of Reasoning: Architecture and Relation to Emotion." *Wiley Reviews: Cognitive Science* 1 (May/June):382–92.

De Waal, F. 1982. *Chimpanzee Politics: Power and Sex Among Apes*. London: Unwin.

Deacon, T. W. 1997. *The Symbolic Species: The Co-evolution of Language and the Brain*. New York: Norton.

Deacon, T. W. 2003. "Universal Grammar and Semiotic Constraints." In *Language Evolution*, edited by M. H. Christiansen and S. Kirby, 111–39. New York: Oxford University Press.

Deane, P. D. 1992. *Grammar in Mind and Brain: Explorations in Cognitive Syntax*. Berlin: Mouton de Gruyter.

Dennett, D. 1992. *Consciousness Explained*. London: Penguin.

Dickinson, E. 1894. *Letters of Emily Dickinson*, vol. 1, edited by M. L. Todd. Boston: Roberts Brothers.

Donald, M. 1991. *Origins of the Modern Mind: Three Stages in the Evolution of Culture and Cognition*. Cambridge, Mass.: Harvard University Press.

Donald, M. 1999. "Preconditions for the Evolution of Protolanguages." In *The Descent of Mind: Psychological Perspectives on Hominid Evolution*, edited by M. C. Corballis and S. E. G. Lea, 138–54. New York: Oxford University Press.

Donald, M. 2001. *A Mind So Rare: The Evolution of Human Consciousness*. New York: Norton.

Donald, M. 2005. "Imitation and Mimesis." In *Imitation, Human Development, and Culture*, edited by S. L. Hurley and N. Chater, 283–300. Perspectives on Imitation from Neuroscience to Social Science, vol. 2. Cambridge, Mass.: MIT Press.

Donald, M. 2007a. "Consciousness and Governance: From Embodiment to Enculturation. An Interview with Merlin Donald and Lars Andreassen." *Cognitive Semiotics* 0:68–83.

Donald, M. 2007b. "The Slow Process: A Hypothetical Cognitive Adaptation for Distributed Cognitive Networks." *Journal of Physiology—Paris* 101:214–22.

Drout, M. D. C. 2006. "A Meme-Based Approach to Oral Traditional Theory." *Oral Tradition* 21(2):269–94.

Dunbar, R. 1996. *Grooming, Gossip and the Evolution of Language.* London: Faber & Faber.

Duncan, R. 1960. *The Opening of the Field.* New York: Grove Press.

Ebbinghaus, H. 1908. *Psychology: An Elementary Text-Book.* Translated by Max Meyer. Boston: D.C. Heath.

Egger, V. 1881. *La Parole intérieure: Essai de psychologie descriptive.* Paris: Librairie Germer Baillière et Cie.

Eliade, M. 1959/1987. *The Sacred and the Profane: The Nature of Religion.* San Diego, Calif.: Harcourt Brace Jovanovich.

Eliot, T. S. 1957. *On Poetry and Poets.* New York: Farrar, Straus and Cudahy.

Emerson, R. W. 2001. *Emerson's Prose and Poetry.* Edited by J. Porte and S. Morris. New York: Norton.

Ericsson, K. A., and W. Kintsch. 1995. "Long-term Working Memory." *Psychological Review* 102:211–45.

Evans, J. St. B. T. 2003. "In Two Minds: Dual-Process Accounts of Reasoning." *Trends in Cognitive Sciences* 7(10):454–59. DOI:10.1016/j.tics.2003.08.012.

Evans, J. St. B. T. 2008. "Dual-Processing Accounts of Reasoning, Judgment, and Social Cognition." *Annual Review of Psychology* 59:255–78.

Evans, J. St. B. T., and K. Frankish, eds. 2009. *In Two Minds: Dual Processes and Beyond.* Oxford: Oxford University Press.

Fagot, J., and J. Vauclair. 1991. "Manual Laterality in Nonhuman Primates: A Distinction Between Handedness and Manual Specialisation." *Psychological Bulletin* 109:76–89.

Fauconnier, G. 1985. *Mental Spaces: Aspects of Meaning Construction in Natural Language.* Cambridge, Mass.: MIT Press.

Fauconnier, G. 1999. "Methods and Generalizations." In *Cognitive Linguistics: Foundations, Scope, and Methodology,* edited by T. Janssen and G. Redeker, 92–127. Berlin: Mouton de Gruyter.

Festinger, L. 1957. *A Theory of Cognitive Dissonance.* Palo Alto, Calif.: Stanford University Press.

Feuer, L. S. 1974. *Einstein and the Generations of Science.* New York: Basic Books.

Fish, S. 1982. *Is There a Text in This Class?* Cambridge, Mass.: Harvard University Press.

Fitch, W. T. 2006. "The Biology and Evolution of Music: A Comparative Perspective." *Cognition* 100:173–215.

Foley, J. M. 1991. *Immanent Art: From Structure to Meaning in Traditional Oral Epic.* Bloomington: Indiana University Press.

Folk, C., and B. Gibson, eds. 2001. *Attraction, Distraction, and Action: Multiple Perspectives on Attentional Capture.* Amsterdam: Elsevier.

Frankish, K. 2010. "Dual-Process and Dual-System Theories of Reasoning." *Philosophy Compass* 5(10):914–26. DOI:10.1111/j.1747-9991.2010.00330.x.

Frith, C. D. 2008. "Social Cognition." In *Philosophical Transactions of the Royal Society B* 363:2033–39.

Galantucci, B., C. A. Fowler, and M. T. Turvey. 2006. "The Motor Theory of Speech Perception Reviewed." *Psychonomic Bulletin & Review* 13(3):361–77.

Gallese, V. 2008. "Il corpo teatrale: Mimetismo, neuroni specchio, simulazione incarnata." *Culture Teatrali* 16:13–38.

Gallese, V., and G. Lakoff. 2005. "The Brain's Concepts: The Role of the Sensory-Motor System in Conceptual Knowledge." *Cognitive Neuropsychology* 22(3/4):455–79.

Gallese, V., L. Fadiga, L. Fogassi, and G. Rizzolatti. 1996. "Action Recognition in the Premotor Cortex." *Brain* 119(2):593–609. DOI:10.1093/brain/119.2.593.

Gazzaniga, M. S., ed. 1994. *The Cognitive Neurosciences*. Cambridge, Mass.: MIT Press.

Genette, G. 1992. *The Architext: An Introduction*. Translated by J. E. Lewin. Berkeley: University of California Press.

Gibson, J. J. 1966. *The Senses Considered as Perceptual Systems*. Boston: Houghton Mifflin.

Gibson, J. J. 1979. *The Ecological Approach to Visual Perception*. Boston: Houghton Mifflin.

Gibson, K. R., and T. Ingold, eds. 1993. *Tools, Language, and Cognition in Human Evolution*. Cambridge, UK: Cambridge University Press.

Gilson, T., and W. Gilson. 2012. *Carved in Stone: The Artistry of Early New England Gravestones*. Middletown, Conn.: Wesleyan University Press.

Girard, R. 1972/1977. *Violence and the Sacred*. Translated by P. Gregory. Baltimore, Md.: Johns Hopkins University Press.

Girard, R. 1978/1987. *Things Hidden Since the Foundation of the World: Research Undertaken in Collaboration with Jean-Michel Oughourlian and G. Lefort*. Stanford, Calif.: Stanford University Press.

Givón, T. 1979. *On Understanding Grammar*. New York: Academic Press.

Givón, T. 1995. *Functionalism and Grammar*. Amsterdam: John Benjamins.

Givón, T. 2002. *Bio-linguistics: The Santa Barbara Lectures*. Amsterdam: John Benjamins.

Givón, T., and B. F. Malle. 2004. *The Evolution of Language Out of Pre-Language*. Amsterdam: John Benjamins.

Glöckner, A., and C. Witteman. 2010. "Beyond Dual-Process Models: A Categorisation of Processes Underlying Intuitive Judgement and Decision Making." *Thinking & Reasoning* 16(1):1–25. DOI:10.1080/13546780903395748.

Goffman, E. 1959. *The Presentation of Self in Everyday Life*. Garden City, N.Y.: Doubleday.

Gomez, A., S. Rousset, and M. Baciu. 2009. "Egocentric-Updating During Navigation Facilitates Episodic Memory Retrieval." *Acta Psychologica* 132:221–27.

Goodale, M. A. 2011. "Transforming Vision into Action." *Vision Research* 51:1567–87.

Goodale, M. A., and A. D. Milner. 1992. "Separate Visual Pathways for Perception and Action." *Trends in Neuroscience* 15:20–25.

Goody, J. 1987. *The Interface Between the Written and the Oral*. Cambridge: Cambridge University Press.

Gould, S. J., and E. Vrba. 1982. "Exaptation—A Missing Term in the Science of Form." *Paleobiology* 8:4–15.
Graff, R. 2001. "Reading and the 'Written Style' in Aristotle's 'Rhetoric.'" *Rhetoric Society Quarterly* 31(4):19–44.
Graves, R. 1948. *The White Goddess.* New York: Vintage Books.
Gray, B. 1971. "Repetition in Oral Literature." *The Journal of American Folklore* 84:289–303.
Green, N., and H. R. Heekeren. 2009. "Perceptual Decision Making: A Bidirectional Link Between Mind and Motion." In *Mind and Motion: The Bidirectional Link Between Thought and Action*, edited by M. Raab et al., 207–18. Progress in Brain Research, vol. 174. DOI:10.1016/S0079-6123(09)01317-X.
Green, R. E., J. Krause, A. W. Briggis, T. Maricic, U. Stenzel, M. Kircher, . . . and S. Pääbo. 2010. "A Draft Sequence of the Neandertal Genome." *Science* 328:710–22. DOI:10.1126/science.1188021.
Guiard, Y. 1987. "Asymmetric Division of Labor in Human Skilled Bimanual Actions: The Kinematic Chain as a Model." *Journal of Motor Behavior* 19:497–517.
Gummere, F. B. 1901. *The Beginnings of Poetry.* New York: Macmillan.
Hall, A. 2008. "The Orality of a Silent Age: The Place of Orality in Medieval Studies." In *Methods and the Medievalist: Current Approaches in Medieval Studies*, edited by M. Lamberg, J. Keskiaho, E. Räsänen, and O. Timofeeva, with L. Virtanen, 270–90. Newcastle upon Tyne, UK: Cambridge Scholars Publishing.
Hardy, A. 1960. "Was Man More Aquatic in the Past?" *New Scientist* 7:642–45.
Harrison, J. E. 1912/1962. *Themis: A Study of the Social Origins of Greek Religion.* New York: Meridien.
Haslam, N., M. Porter, and L. Rothschild. 2001. "Visual Search: Efficiency Continuum or Distinct Processes?" *Psychonomic Bulletin & Review* 8(4):742–46.
Havelock, E. A. 1963. *Preface to Plato.* Cambridge, Mass.: Belknap Press of Harvard University Press.
Havelock, E. A. 1986. *The Muse Learns to Write: Reflections on Orality and Literacy from Antiquity to the Present.* New Haven: Yale University Press.
Heidegger, M. 1927/1962. *Being and Time.* Translated by J. Macquarrie, and E. Robinson. New York: Harper & Row.
Heinrichs, A. 1998. "Dromena und Legomena: Zum Rituellen Selbstverständnis der Griechen." In *Ansichten Griechischer Rituale: Geburtstags-Symposium für Walter Burkert*, edited by W. Burkert and F. Graf, 33–71. Stuttgart: B. G. Teubner.
Helmholtz, H. von. 1866/2000. *Helmholtz's Treatise on Physiological Optics.* Translated by J. P. C. Southall. New York: Thoemmes Continuum.
Henriques, G. 2003. "The Tree of Knowledge System and the Theoretical Unification of Psychology." *Review of General Psychology* 7(2):150–82.
Hesslow, G. 2002. "Conscious Thought as Simulation of Behaviour and Perception." *Trends in Cognitive Science* 6:242–47.
Hewes, G. W. 1973. "Primate Communication and the Gestural Origin of Language." *Current Anthropology* 14:5–12.

Hickok, G., and D. Poeppel. 2004. "Dorsal and Ventral Streams: A Framework for Understanding Aspects of the Functional Anatomy of Language." *Cognition* 92:67–99.

Hipp, J. F., D. J. Hawellek, M. Corbetta, M. Siegel, and A. K. Engel. 2012. "Large-Scale Cortical Correlation Structure of Spontaneous Oscillatory Activity." *Nature Neuroscience* 15:884–90. DOI:10.1038/nn.3101.

Hockett, C. F. 1960/1982. "The Origin of Speech." Originally printed in *Scientific American* 293(88). Reprinted in *Human Communication: Language and Its Psychobiological Bases. Readings from Scientific American*, edited by Shih Yi Wang, 5–12. San Francisco: W. H. Freeman.

Hudson, R. 2001. Review of Terrence Deacon, *The Symbolic Species: The Co-evolution of Language and the Human Brain*. *Journal of Pragmatics* 33:129–35.

Huizinga, J. 1938/1970. *Homo Ludens: A Study of the Play Element in Culture*. London: Paladin.

Hunt, K. D. 1994. "The Evolution of Human Bipedality: Ecology and Functional Morphology." *Journal of Human Evolution* 26:182–202.

Hurford, J. R. 2003. "The Neural Basis of Predicate-Argument Structure." *Behavioral and Brain Sciences* 26(3):261–316.

Hurford, J. R. 2007. "The Origin of Noun Phrases: Reference, Truth, and Communication." *Lingua* 117:527–42.

Hurford, J. R., M. Studdert-Kennedy, and C. Knight, eds. 1998. *Approaches to the Evolution of Language: Social and Cognitive Bases*. Cambridge, UK: Cambridge University Press.

Hurley, S. L. 1998. *Consciousness in Action*. Cambridge, Mass.: Harvard University Press.

Hurley, S. L. 2005. "The Shared Circuits Hypothesis: A Unified Functional Architecture for Control, Imitation, and Simulation." In *Perspectives on Imitation from Neuroscience to Social Science*, vol. 1: *Mechanisms of Imitation and Imitation in Animals*, edited by S. L. Hurley and N. Chater, 177–94. Cambridge, Mass.: MIT Press.

Hurley, S. L., and N. Chater. 2005a. *Perspectives on Imitation from Neuroscience to Social Science*. Vol. 1: *Mechanisms of Imitation and Imitation in Animals*. Cambridge, Mass.: MIT Press.

Hurley, S. L., and N. Chater, eds. 2005b. *Perspectives on Imitation from Neuroscience to Social Science*. Vol. 2: *Imitation, Human Development, and Culture*. Cambridge, Mass.: MIT Press.

Iacoboni, M. 2005. "Understanding Others: Imitation, Language, and Empathy." In *Perspectives on Imitation from Neuroscience to Social Science*, Volume 1, 77–99. Cambridge, Mass.: MIT Press.

Ingle, D. J., M. A. Goodale, and R. J. W. Mansfield, eds. 1982. *Analysis in Visual Behavior*. Cambridge, Mass.: MIT Press.

Jackendoff, R. S. 1987. *Consciousness and the Computational Mind*. Cambridge, Mass.: MIT Press.

Jackson, T. 2000. "Questioning Interdisciplinarity: Cognitive Science, Evolutionary Psychology, and Literary Criticism." *Poetics Today* 21:319–47.

Jackson, T. 2002. "Issues and Problems in the Blending of Cognitive Science, Evolutionary Psychology, and Literary Study." *Poetics Today* 23:161–79.

Jackson, T. 2003. "'Literary Interpretation' and Cognitive Literary Studies." *Poetics Today* 24:191–205.

Jakobson, R. 1960. "Closing Statement: Linguistics and Poetics." In *Style in Language*, edited by T. A. Sebeok, 350–77. New York: Wiley.

Jakobson, R., and M. Halle. 1956. *Fundamentals of Language*. The Hague: Mouton.

James, W. 1890/1950. *The Principles of Psychology*, vol. 1. New York: Dover.

Jamison, S. W., H. C. Melchert, and B. Vine, eds. 2010. *Proceedings of the 21st Annual UCLA Indo-European Conference*. Bremen, Germany: Hempen.

Janssen, T., and G. Redeker. 1999. *Cognitive Linguistics: Foundations, Scope, and Methodology*. Berlin: Mouton de Gruyter.

Jeannerod, M. 1997. *The Cognitive Neuroscience of Action*. Oxford, UK: Blackwell.

Jeannerod, M. 2006. *Motor Cognition: What Actions Tell the Self*. Oxford, UK: Oxford University Press.

Jeannerod, M., and P. Jacob. 2005. "Visual Cognition: A New Look at the Two-Visual Systems Model." *Neuropsychologia* 43:301–10.

Jolles, A. 1930. *Einfache Formen: Legende, Sage, Mythe, Rätsel, Spruch, Kasus, Memorabile, Märchen, Witz*. Halle (Saale), Germany: M. Niemeyer.

Kaas, J. H., and C. E. Collins, eds. 2003. *The Primate Visual System*. London: Taylor & Francis (CRC Press).

Kaas, J. H., and T. A. Hackett. 1999. "'What' and 'Where' Processing in Auditory Cortex." *Nature Neuroscience* 2:1045–47.

Kelly, R. E. 2001. "Tripedal Knuckle-Walking: A Proposal for the Evolution of Human Locomotion and Handedness." *Journal of Theoretical Biology* 213:333–58.

Kemmerer, D. 2010. "How Words Capture Visual Experience: The Perspective from Cognitive Neuroscience." In *Words and the Mind: How Words Capture Human Experience*, edited by B. Malt and P. Wolff, 287–327. Oxford: Oxford University Press.

Kennedy, G. A. 1992. "A Hoot in the Dark: The Evolution of General Rhetoric." *Philosophy & Rhetoric* 25(1):1–21.

Keren, G., and Y. Schul. 2009. "Two Is Not Always Better Than One: A Critical Evaluation of Two-System Theories." *Perspectives on Psychological Science* 4(6):533–50.

Klein, R. M., D. P. Munoz, M. C. Dorris, and T. L. Taylor. 2001. "Inhibition of Return in Monkey and Man." In *Attraction, Distraction, and Action: Multiple Perspectives on Attentional Capture*, edited by C. Folk and B. Gibson, 27–47. Amsterdam: Elsevier.

Knight, C. 1998. "Ritual/Speech Coevolution: A Solution to the Problem of Deception." In *Approaches to the Evolution of Language: Social and Cognitive Bases*, edited by J. R. Hurford, M. Studdert-Kennedy, and C. Knight, 68–91. Cambridge, UK: Cambridge University Press.

Knight, C. 2000. "Play as Precursor of Phonology and Syntax." In *The Evolutionary Emergence of Language: Social Function and the Origins of Linguistic Form*, edited by C. Knight, M. Studdert-Kennedy, and J. R. Hurford, 99–119. Cambridge, UK: Cambridge University Press.

Knight, C., M. Studdert-Kennedy, and J. R. Hurford. (2000). "Language: A Darwinian Adaptation?" In *The Evolutionary Emergence of Language: Social Function and the Origins of Linguistic Form*, edited by C. Knight, M. Studdert-Kennedy, and J. R. Hurford, 1–15. Cambridge, UK: Cambridge University Press.

Kobayashi, H., and S. Kohshima. 2001. "Unique Morphology of the Human Eye and Its Adaptive Meaning: Comparative Studies on External Morphology of the Primate Eye." *Journal of Human Evolution* 40:419–35.

Koestler, A. 1967/1990. *The Ghost in the Machine.* New York: Penguin.

Kosslyn, S. M. 1980. *Image and Mind.* Cambridge, Mass.: Harvard University Press.

Kosslyn, S. M., and A. L. Sussman. 1994. "Roles of Imagery in Perception: Or, There Is No Such Thing as Immaculate Perception." In *The Cognitive Neurosciences*, edited by M. S. Gazzaniga, 1035–42. Cambridge, Mass.: MIT Press.

Kosslyn, S. M., W. J. Thompson, and G. Ganis. 2006. *The Case for Mental Imagery.* New York: Oxford University Press.

Kramár, E. A., A. H. Babayan, C. F. Gavin, C. D. Cox, M. Jafari, C. M. Gall, G. Rumbaugh, and G. Lynch. 2012. "Synaptic Evidence for the Efficacy of Spaced Learning." *Proceedings of the National Academy of Sciences,* March 12, 2012 (published online before print, DOI:10.1073/pnas.1207001090).

Kramnick, J. 2011. "Against Literary Darwinism." *Critical Inquiry* 37(2):315–47.

Kramnick, J. 2012. "Literary Studies and Science: A Reply to My Critics." *Critical Inquiry* 38(2):431–60.

Krifka, M. 2007. "Functional Similarities Between Bimanual Coordination and Topic/Comment Structure." In *Working Papers of the SFB 632, Interdisciplinary Studies on Information Structure (ISIS)*, edited by S. Ishihara, S. Jannedy, and A. Schwarz, 39–59. Potsdam: Universitätsverlag Potsdam. (Also in *Variation, Selection, Development. Probing the Evolutionary Model of Language Change*, edited by R. Eckardt, G. Jäger, and T. Veenstra, 307–36. Berlin/New York: Mouton de Gruyter, 2008.)

Kristeva, J., ed. 1971. *Essays in Semiotics.* The Hague: Mouton.

Lakoff, G. 1987. *Women, Fire, and Dangerous Things: What Categories Reveal About the Mind.* Chicago: University of Chicago Press.

Lakoff, G., and M. Johnson. 1980. *Metaphors We Live By.* Chicago: University of Chicago Press.

Lakoff, G., and M. Turner. 1989. *More than Cool Reason: A Field Guide to Poetic Metaphor.* Chicago: University of Chicago Press.

Lancy, D. F. 1980. "Play in Species Adaptation." *Annual Review of Anthropology* 9:471–95.

Landau, B., and R. Jackendoff. 1993. "'What' and 'Where' in Spatial Language and Spatial Cognition." *Behavioral and Brain Sciences* 16:217–65.

Langacker, R. W. 1987. *Foundations of Cognitive Grammar.* Vol. 1, *Theoretical Prerequisites.* Stanford, Calif: Stanford University Press.

Langacker, R. W. 1999. *Grammar and Conceptualization.* Berlin: Mouton de Gruyter.

Langacker, R. W. 2001. "Dynamicity in Grammar." *Axiomathes* 12:7–33.

Langacker, R. W. 2002. *Concept, Image, and Symbol: The Cognitive Basis of Grammar.* Berlin: Mouton de Gruyter.

Langer, S. 1953. *Feeling and Form*. New York: Scribner.
Leakey, L. S. B., P. V. Tobias, and J. R. Napier. 1964. "A New Species of the Genus *Homo* from Olduvai Gorge." *Nature* 202:7–9.
Leslie, A. M. 1987. "Pretense and Representation: The Origins of 'Theory of Mind.'" *Psychological Review* 94(4):412–26.
Leslie, A. M. 2002. "Pretense and Representation Revisited." In *Representation, Memory, and Development*, edited by N. L. Stein, P. J. Bauer, and M. Rabinowitz, 103–14. Mahwah, N.J.: Lawrence Erlbaum Associates.
Levelt, W. J. M. 1989. *Speaking: From Intention to Articulation*. Cambridge, Mass.: MIT Press.
Levelt, W. J. M. 1992. "Accessing Words in Speech Production: Stages, Processes and Representations." *Cognition* 42:1–22.
Liberman, A. M., and I. G. Mattingly. 1985. "The Motor Theory of Speech Perception Revised." *Cognition* 21:1–36.
Liberman, A. M., and D. H. Whalen. 2000. "On the Relation of Speech to Language." *Trends in Cognitive Sciences* 4:187–96.
Lieberman, P. 1984. *The Biology and Evolution of Language*. Cambridge, Mass.: Harvard University Press.
Lifter, K., and L. Bloom. 1989. "Object Knowledge and the Emergence of Language." *Infant Behavior and Development* 12:385–423.
Lin, P. M. S. 2010. "The Phonology of Formulaic Sequences: A Review." In *Perspectives on Formulaic Language: Acquisition and Communication*, edited by D. Wood, 174–93. London: Continuum.
Luria, A. R. 1987. *The Mind of a Mnemonist: A Little Book About a Vast Memory*. Cambridge, Mass.: Harvard University Press.
MacKenzie, C. L., and T. Iberall. 1994. *The Grasping Hand*. Amsterdam: Elsevier.
MacKenzie, I. 2000. "Improvisation, Creativity, and Formulaic Language." *The Journal of Aesthetics and Art Criticism* 58(2):173–79.
MacNeilage, P. F. 1999. "Whatever Happened to Articulate Speech?" In *The Descent of Mind: Psychological Perspectives on Hominid Evolution*, edited by M. C. Corballis and S. E. G. Lea, 116–37. New York: Oxford University Press.
MacWhinney, B. 2004. "The Gradual Emergence of Language." In *The Evolution of Language Out of Pre-Language*, edited by Givón and Malle, 233–63. Amsterdam: John Benjamins.
Malt, B. C., and P. M. Wolff. 2010. *Words and the Mind: How Words Capture Human Experience*. New York: Oxford University Press.
Mameli, M. 2001. "Mindreading, Mindshaping, and Evolution." *Biology and Philosophy* 16:597–628.
Mandler, G. 1985. *Cognitive Psychology: An Essay in Cognitive Science*. Hillsdale, N.J.: L. Erlbaum Associates.
Marean, C. W., M. Bar-Matthews, J. Bernatchez, E. Fisher, P. Goldberg, A. Herries, . . . and H. M. Williams. 2007. "Early Human Use of Marine Resources and Pigment in South Africa During the Middle Pleistocene." *Nature* 449:905–8.
Martinet, A., and L. R. Palmer. 1964. *Elements of General Linguistics*. Chicago: University of Chicago Press.

Marzke, M. W., and R. F. Marzke. 2000. "Evolution of the Human Hand: Approaches to Acquiring, Analyzing, and Interpreting the Anatomical Evidence." *Journal of Anatomy* 196:121–40.

Matlock, T. 2004. "Fictive Motion as Cognitive Simulation." *Memory & Cognition* 32:1389–1400.

McBride, G. 1971. "On the Evolution of Human Language." In *Essays in Semiotics*, edited by J. Kristeva, 560–68. The Hague: Mouton.

McCune-Nicolich, L. 1981. "Toward Symbolic Functioning: Structure of Early Pretend Games and Potential Parallels with Language." *Child Development* 52(3):785–97.

McGrath, T. 1982. "Language, Power, and Dream." In *Claims for Poetry*, edited by D. Hall, 286–95. Ann Arbor: University of Michigan Press.

McGurk, H., and J. McDonald. 1976. "Hearing Lips and Seeing Voices." *Nature* 264:746–48.

McLuhan, M. 1962/2011. *The Gutenberg Galaxy: The Making of Typographic Man*. Toronto: University of Toronto Press.

McLuhan, M. 1964. *Understanding Media: The Extensions of Man*. London: Routledge and Kegan Paul.

McNeill, D. 1992. *Hand and Mind: What Gestures Reveal About Thought*. Chicago: University of Chicago Press.

McNeill, D., S. D. Duncan, J. Cole, S. Gallagher, and B. Bertenthal. 2010. "Growth Points from the Very Beginning." In *The Emergence of Protolanguage: Holophrasis vs. Compositionality*, edited by M. Arbib and D. Bickerton, 117–32. Amsterdam: John Benjamins.

Merker, B. H., G. S. Madison, and P. Eckerdal. 2009. "On the Role and Origin of Isochrony in Human Rhythmic Entrainment." *Cortex* 45:4–17.

Merleau-Ponty, M. 1945/1962. *The Phenomenology of Perception*. Translated by C. Smith. London: Routledge and Kegan Paul.

Meyer, M. W. 1999. *The Ancient Mysteries: A Sourcebook of Sacred Texts*. Philadelphia: University of Pennsylvania Press.

Miall, D. S. 2005. "Beyond Interpretation: The Cognitive Significance of Reading." In *Cognition and Literary Interpretation in Practice*, edited by H. Veivo, B. Pettersson, and M. Polvinen, 129–56. Helsinki: University of Helsinki Press.

Miall, D. S. 2006. *Literary Reading: Empirical and Theoretical Studies*. New York: Peter Lang.

Miller, G. A., 1956. "The Magical Number Seven, Plus or Minus Two: Some Limits on Our Capacity for Processing Information." *Psychological Review* 63:81–97.

Milner, A. D., and M. A. Goodale. 1995. *The Visual Brain in Action*. New York: Oxford University Press.

Milner, A. D., and M. A. Goodale. 2008. "Two Visual Systems Re-Viewed." *Neuropsychologia* 45:774–85.

Mitchell, W. J. T. 2005. "There Are No Visual Media." *Journal of Visual Culture* 4:257–66.

Mithen, S. J. 1996. *The Prehistory of the Mind: A Search for the Origins of Art, Religion, and Science*. London: Thames and Hudson.

Mithen, S. J. 2006. *The Singing Neanderthals: The Origins of Music, Language, Mind, and Body*. Cambridge, Mass.: Harvard University Press.

Monaghan, P., M. H. Christiansen, and S. A. Fitneva. 2011. "The Arbitrariness of the Sign: Learning Advantages from the Structure of the Vocabulary." *Journal of Experimental Psychology: General* 140:325–47.

Mühlhäusler, P., and R. Harré. 1990. *Pronouns and People: The Linguistic Construction of Social and Personal Identity*. Oxford: Blackwell.

Müller, F. M. 1868. *Lectures on the Science of Language Delivered at the Royal Institution of Great Britain . . . 1861 [and 1863]*. New York: Scribner.

Müller, F. M. 1889. *Three Lectures on the Science of Language and Its Place in General Education* [delivered at the Oxford University Extension Meeting 1889]. London: Longmans, Green.

Murray, G. 1927/1957. *The Classical Tradition in Poetry*. New York: Vintage Books.

Napier, J. R. 1956. "The Prehensile Movements of the Human Hand." *Journal of Bone and Surgery* 38:902–13.

Napier, J. R. 1980. *Hands*. New York: Pantheon.

Neisser, U. 1976. *Cognition and Reality: Principles and Implications of Cognitive Psychology*. San Francisco: W. H. Freeman.

Nietzsche, F. W., R. Geuss, and R. Speirs. 1872/1999. *The Birth of Tragedy and Other Writings*. Cambridge Texts in the History of Philosophy. Cambridge, UK: Cambridge University Press.

Noë, A. 2004. *Action in Perception*. Cambridge, Mass.: MIT Press.

Norenzayan, A., S. Atran, J. Faulkner, and J. M. Schaller. 2006. "Memory and Mystery: The Cultural Selection of Minimally Counterintuitive Narratives." *Cognitive Science* 30:531–53.

Norman, J. 2002. "Two Visual Systems and Two Theories of Perception: An Attempt to Reconcile the Constructivist and Ecological Approaches." *Behavioral and Brain Sciences* 25:73–144.

Nöth, W. 1995. *Handbook of Semiotics*. Bloomington: Indiana University Press.

Oliveira, R. F. de, L. Damisch, E.-J. Hossner, R. R. R. Oudejans, M. Raab, K. G. Volz, and A. M. Williams. 2009. "The Bidirectional Links Between Decision Making, Perception, and Action." In *Mind and Motion: The Bidirectional Link Between Thought and Action*, edited by M. Raab et al., 85–93. Brain Research 174. DOI:10.1016/S0079-6123(09)01308-9.

Olson, C. 1950/1959. *Projective Verse*. New York: Totem Press.

Ong, W. J. 1982. *Orality and Literacy: The Technologizing of the Word*. New York: Methuen.

Paivio, A. 1971. *Imagery and Verbal Processes*. New York: Holt, Rinehart & Winston.

Paivio, A. 1990. *Mental Representations: A Dual Coding Approach*. New York: Oxford University Press.

Paivio, A. 2007. *Mind and Its Evolution: A Dual Coding Theoretical Approach*. Mahwah, N.J.: Lawrence Erlbaum Associates.

Parker, G. J. M., S. Luzzi, D. C. Alexander, C. A. M. Wheeler-Kingshott, O. Ciccarelli, and M. A. L. Ralph. 2005. "Lateralization of Ventral and Dorsal Auditory-Language Pathways in the Human Brain." *NeuroImage* 24:656–66.

Pashler, H. 1999. *The Psychology of Attentioon*. Cambridge, Mass.: MIT Press.
Patel, A. D. 2008. *Music, Language, and the Brain*. New York: Oxford University Press.
Patel, A. D. 2011. "Why Would Musical Training Benefit the Neural Encoding of Speech? The OPERA Hypothesis." *Frontiers in Psychology* 2(142):1–14. DOI:10.3389/fpsyg.2011.00142.
Patkin, M. 1965. "The Hand Has Two Grips: An Aspect of Surgical Dexterity." *The Lancet* 283:1384–85. DOI:10.1016/S0140-6736(64)91680-0.
Peirce, C. S. 1931. *The Collected Papers of Charles Sanders Peirce*, vol. 2. Edited by C. Hartshorne and P. Weiss. Cambridge, Mass.: Harvard University Press.
Pellegrini, A. D., and P. K. Smith, eds. 2004. *The Nature of Play: Great Apes and Humans*. New York: Guilford Press.
Piantadosi, S. T., H. Tily, and E. Gibson. 2012. "The Communicative Function of Ambiguity in Language." *Cognition* 122(3):280–91.
Pinker, S. 1994. *The Language Instinct*. New York: William Morrow.
Polanyi, M. 1958. *Personal Knowledge: Toward a Post-Critical Philosophy*. Chicago: University of Chicago Press.
Pontzer, H., J. H. Holloway, D. A. Raichlen, and D. E. Lieberman. 2009. "Control and Function of Arm Swing in Human Walking and Running." *Journal of Experimental Biology* 212:523–34.
Posner, M. I. 1978. *Chronometric Explorations of Mind*. Hillsdale, N.J.: Erlbaum.
Posner, M. I. 1980. "Orienting of Attention." *Quarterly Journal of Experimental Psychology* 32:3–25.
Posner, M. I., R. D. Rafal, L. S. Choate, and J. Vaughan. 1985. "Inhibition of Return: Neural Basis and Function." *Cognitive Neuropsychology* 2:211–28.
Pound, E., and E. F. Fenollosa. 1920. *Instigations: Together with an Essay on the Chinese Written Character*. New York: Boni and Liveright.
Pound, L. 1917. "The Beginnings of Poetry." *Publications of the Modern Language Association of America* 32(2):201–32.
Prinz, J. 2006. "Putting the Brakes on Enactive Perception." *Psyche* 12. http://psyche.cs.monash.edu.au/.
Prinz, W. 1983. "Modes of Linkage Between Perception and Action." In *Cognition and Motor Processes*, edited by W. Prinz and A. F. Sanders, 185–93. Berlin: Springer.
Prinz, W., and A. F. Sanders, eds. 1983. *Cognition and Motor Processes*. Berlin: Springer.
Ramsey, J. K., and W. C. McGrew. 2004. "Object Play in Great Apes." In *The Nature of Play: Great Apes and Humans*, edited by A. D. Pellegrini and P. K. Smith, 89–112. New York: Guilford Press.
Rauschecker, J. P. 1998. "Cortical Processing of Complex Sounds." *Current Opinion in Neurobiology* 8:516–21.
Redford, J. S. 2010. "Evidence of Metacognitive Control by Humans and Monkeys in a Perceptual Categorization Task." *Journal of Experimental Psychology: Learning, Memory, and Cognition* 36(1):248–54. DOI:10.1037/a0017.
Regier, T., and P. Kay. 2009. "Language, Thought, and Color: Whorf Was Half Right." *Trends in Cognitive Sciences* 13(10):439–46. DOI:10.1016/j.tics.2009.07.001.

Richardson, A., and F. F. Steen, eds. 2002. *Literature and the Cognitive Revolution*. Special Issue. *Poetics Today* 23.

Richardson, D., and T. Matlock. 2007. "The Integration of Figurative Language and Static Depictions: An Eye Movement Study of Fictive Motion." *Cognition* 102:129–38.

Ricoeur, P. 1991. *From Text to Action: Essays in Hermeneutics, II*. Evanston, Ill.: Northwestern University Press.

Rilke, R. M. 1922/1965. In *Rainer Maria Rilke's Duineser Elegien*, edited by E. L. Stahl. Oxford: B. Blackwell.

Romanski, L. M., B. Tian, J. Fritz, M. Mishkin, P. S. Goldman-Rakic, and J. P. Rauschecker. 1999. "Dual Streams of Auditory Afferents Target Multiple Domains in the Primate Prefrontal Cortex." *Nature Neuroscience* 2:1131–36.

Rosenberg, B. 1970. "The Formulaic Quality of Spontaneous Sermons." *The Journal of American Folklore* 83:327, 3–20.

Rousseau, J.-J. 1754/1984. *Discourse on the Origin of Inequality Among Men*. Translated by G. D. H. Cole. Chicago: Encyclopaedia Britannica.

Rubin, D. C. 1995. *Memory in Oral Traditions: The Cognitive Psychology of Epic, Ballads, and Counting-Out Rhymes*. New York: Oxford University Press.

Rudnytsky, P. L., ed. 1993. *Transitional Objects and Potential Spaces: Literary Uses of D. W. Winnicott*. New York: Columbia University Press.

Russo, J. 1994. "Homer's Style: Nonformulaic Features of an Oral Aesthetic." *Oral Tradition* 9(2):371–89.

Sagan, C., and A. Druyan. 1992. *Shadows of Forgotten Ancestors: A Search for Who We Are*. New York: Ballantine.

Samuels, R. 2009. "The Magical Number Two, Plus or Minus: Dual Process Theory as a Theory of Cognitive Kinds." In *In Two Minds: Dual Process and Beyond*, edited by J. Evans and K. Frankish, 129–46. New York: Oxford University Press.

Saygin, A. P., S. McCullough, M. Alac, and K. Emmorey. 2010. "Modulation of BOLD Response in Motion-Sensitive Lateral Temporal Cortex by Real and Fictive Motion Sentences." *Journal of Cognitive Neuroscience* 22(11):2480–90.

Schechner, R. 1985. *Between Theater and Anthropology*. Philadelphia: University of Pennsylvania Press.

Scherer, K. R. 2001. "Appraisal Considered as a Process of Multilevel Sequential Checking." In *Appraisal Processes in Emotion Theory, Methods, and Research*, edited by K. R. Scherer, A. Schorr, and T. Johnstone, 92–120. New York: Oxford University Press.

Scherer, K. R., A. Schorr, and T. Johnstone, eds. 2001. *Appraisal Processes in Emotion Theory, Methods and Research*. New York: Oxford University Press.

Schick, K. D., and N. Toth. 1994. *Making Silent Stones Speak: Human Evolution and the Dawn of Technology*. New York: Simon & Schuster.

Schoonheim, P. L. 2000. *Aristotle's Meteorology in the Arabico-Latin Tradition*. Boston: Brill.

Schwartz, G. E., R. J. Davidson, and D. Shapiro, eds. 1986. *Consciousness and Self-Regulation. Advances in Research and Theory*, vol. 4. New York: Plenum Press.

Sherrington, C. S. 1940. *Man on His Nature.* Cambridge, UK: Cambridge University Press.

Sebeok, T. A., ed. 1960. *Style in Language.* New York: Wiley.

Segal, C. 1999. *Tragedy and Civilization: An Interpretation of Sophocles.* Norman: University of Oklahoma Press.

Silverman, K. 1983. *The Subject of Semiotics.* New York: Oxford University Press.

Siviy, S. M. 1998. "Neurobiological Substrates of Play Behavior: Glimpses into the Structure and Function of Mammalian Playfulness." In *Animal Play: Evolutionary, Comparative, and Ecological Perspectives,* edited by M. Bekoff and J. A. Byers, 221–42. Cambridge, UK: Cambridge University Press.

Sloman, S. A. 1996. "The Empirical Case for Two Systems of Reasoning." *Psychological Bulletin* 119(1):3–22.

Smail, D. L. 2008. *On Deep History and the Brain.* Berkeley: University of California Press.

Small, J. P. 1997. *Wax Tablets of the Mind: Cognitive Studies of Memory and Literacy in Classical Antiquity.* London: Routledge.

Sokolov, A. N. 1972. *Inner Speech and Thought.* Monographs in Psychology. New York: Plenum Press.

Sontag, S. 1966/2001. *Against Interpretation, and Other Essays.* New York: Picador.

Sophocles. 1939. *The Oedipus Cycle: An English Version.* Translated by D. Fitts and R. Fitzgertald. New York: Harcourt Brace Jovanovich.

Sperry, R. W. 1952. "Neurology and the Mind-Body Problem." *American Scientist* 40:291–312.

Stamenov, M., and V. Gallese, eds. 2002. *Mirror Neurons and the Evolution of Brain and Language.* Advances in Consciousness Research 42. Amsterdam: John Benjamins.

Stanovich, K. E., and R. F. West. 2003. "Evolutionary Versus Instrumental Goals: How Evolutionary Psychology Misconceives Human Rationality." In *Evolution and the Psychology of Thinking: The Debate,* edited by D. Over, 171–230. Hove, UK: Psychology Press.

Stein, N. L., P. J. Bauer, and M. Rabinowitz, eds. 2002. *Representation, Memory, and Development.* Essays in Honor of Jean Mandler. Mahwah, N.J.: Lawrence Erlbaum Associates.

Sterelny, K. 2007. "Social Intelligence, Human Intelligence and Niche Construction." In *Philosophical Transactions of the Royal Society B* 362:719–30.

Stockwell, P. 2002. *Cognitive Poetics: An Introduction.* London: Routledge.

Studdert-Kennedy, M. 1991. "Leap of Faith: A Review of Derek Bickerton's *Language and Species.*" *Haskins Laboratories Status Report on Speech Research,* SR·107/108:255–62.

Suddendorf, T. 1999. "The Rise of the Metamind." In *The Descent of Mind: Psychological Perspectives on Hominid Evolution,* edited by M. C. Corballis and S. E. G. Lea, 218–60. New York: Oxford University Press.

Svenbro, J. 1976. *La Parole et le marbre: Aux origines de la poétique grecque.* Lund, Sweden: J. Svenbro.

Svenbro, J. 1993. *Phrasikleia: An Anthropology of Reading in Ancient Greece*. Ithaca, N.Y.: Cornell University Press.
Talmy, L. 2000. *Toward a Cognitive Semantics*. Cambridge, Mass.: MIT Press.
Thoreau, H. D. 1962. *The Journal of Henry David Thoreau*, Vol. 1 *(1837–October 1855)*, edited by B. Torrey and F. H. Allen. New York: Dover.
Tomasello, M. 2009. *Why We Cooperate*. Cambridge, Mass.: MIT Press.
Tooby, J., and L. Cosmides. 1992. "The Psychological Foundations of Culture." In *The Adapted Mind*, edited by H. Barkow, L. Cosmides, and J. Tooby, 19–136. New York: Oxford University Press.
Treisman, A. M., and G. Gelade. 1980. "A Feature-Integration Theory of Attention." *Cognitive Psychology* 12:97–136.
Tulving, E. 1983. *Elements of Episodic Memory*. New York: Oxford University Press.
Tulving, E. 2002. "Episodic Memory: From Mind to Brain." *Annual Review of Psychology* 53:1–25.
Turner, M. 1991. *Reading Minds: The Study of English in the Age of Cognitive Science*. Princeton, N.J.: Princeton University Press.
Turner, M. 2002. "Review: *Toward a Cognitive Semantics* by Leonard Talmy." *Language* 78(3):576–78.
Turner, V. 1969. *Ritual Process: Structure and Anti-Structure*. Chicago: Aldine.
Uexküll, J. von. 1921. *Umwelt und Innenwelt der Thiere*. Berlin: Springer.
Ungerleider, L. G., and M. Mishkin. 1982. "Two Cortical Visual Systems: Separation of Appearance and Location of Objects." In *Analysis in Visual Behavior*, edited by D. J. Ingle, M. A. Goodale, and R. J. W. Mansfield, 549–86. Cambridge, Mass.: MIT Press.
Van Gennep, A. 1909/1966. *Rites of Passage*. Chicago: University of Chicago Press.
Vermeule, B. 2012. "Wit and Poetry and Pope, or The Handicap Principle." *Critical Inquiry* 38(2):426–30.
Wade, N. July 8, 2010. "Clues of Britain's First Humans." *New York Times*, A6.
Wallentin, M., S. Østergaard, T. E. Lund, L. Østergaard, and A. Roepstorff. 2005. "Concrete Spatial Language: See What I Mean?" *Brain and Language* 92:221–33.
Wang, R. F., and E. S. Spelke. 2002. "Human Spatial Representation: Insights from Animals." *Trends in Cognitive Sciences* 6:376–82. PII:S1364–6613(02)01961.
Welsh, A. 1987. *Roots of Lyric: Primitive Poetry and Modern Poetics*. Princeton N.J.: Princeton University Press.
Williams, W. C. 1966. *The William Carlos Williams Reader*, edited by M. L. Rosenthal. New York: Published for J. Laughlin by New Directions Pub. Corp.
Wilson, E. O. 1975. *Sociobiology: The New Synthesis*. Cambridge, Mass.: Belknap Press of Harvard University Press.
Winnicott, D. W. 1982/2005. *Playing and Reality*. New York: Tavistock/Routledge.
Wolfe, J. M. 1998. "What Can 1 Million Trials Tell Us About Visual Search?" *Psychological Science* 9:33–39.
Wood, D. 2010. *Perspectives on Formulaic Language: Acquisition and Communication*. London: Continuum.
Worthen, J. B., and R. R. Hunt. 2008. "Mnemonics: Underlying Processes and Practical Applications." *Learning and Memory: A Comprehensive Reference*, 145–56.

Wray, A. 1998. "Protolanguage as a Holistic System for Social Interaction." *Language & Communication* 18:47–67.
Wray, A. 2002a. *Formulaic Language and the Lexicon*. Cambridge, UK: Cambridge University Press.
Wray, A., ed. 2002b. *Transition to Language: Studies in the Evolution of Language*. New York: Oxford University Press.
Wray, A., and G. W. Grace. 2007. "The Consequences of Talking to Strangers: Evolutionary Corollaries of Socio-Cultural Influences on Linguistic Form." *Lingua* 117:543–78.
Wray, A., and M. R. Perkins. 2000. "The Functions of Formulaic Language: An Integrated Model." *Language & Communication* 20:1–28.
Wright, R. D., and L. M. Ward. 2008. *Orienting of Attention: Orienting Attentional Focus to Locations and Objects in Visual Space*. New York: Oxford University Press.
Wundt, W. 1900/1970. "The Psychology of the Sentence." In *Language and Psychology: Historical Aspects of Psycholinguistics*, edited and translated by A. L. Blumenthal, 20–31. New York: Wiley.
Yates, F. A. 1966. *The Art of Memory*. London: Routledge and K. Paul.
Young, R. W. 2003. "Evolution of the Human Hand: The Role of Throwing and Clubbing." *Journal of Anatomy* 202:165–74.
Zlatev, J. 2008. "From Proto-Mimesis to Language: Evidence from Primatology and Social Neuroscience." *Journal of Physiology—Paris* 102:137–51.
Zumthor, P. 1990. *Oral Poetry: An Introduction*. Translated by Kathryn Murphy-Judy. Minneapolis: University of Minnesota Press.
Zwaan, R. A., and M. P. Kaschak. 2006. "Language, Visual Cognition and Motor Action." In *Encyclopedia of Language and Linguistics*, 2nd ed., vol. 6, edited by K. Brown, 648–51. Oxford: Elsevier.

Index

abbreviation, semiotic, 116, 128–29, 136, 147, 184, 221n4. *See also* emblem; gesture
Acheulian culture, 115, 147
allocentric. *See* spatial frames of reference
apostrophe, 170, 212
Arbib, Michael A., 47, 136–37, 181, 224n8
Aristotle: hermeneutics, 13, 216n8; poetics, 2–3, 12, 63, 184, 211–12; rhetoric, 153, 170, 188, 192, 208–9
auditory perception, 22, 210, 222n6; brain, 130–34, 199; icons, 115, 134; symbols, 126, 185, 201, 214
Australopithecus, 66, 75, 97, 109, 111, 114, 121

Barsalou, Lawrence, 52, 91–92
Bateson, Gregory, 20, 65–66, 78
Belin, Pascal, 133
Bell, David, 145–46
Benveniste, Emile, 78, 145, 183
Bickerton, Derek, 23, 113–17, 125, 137, 221n2
big history, 8–9
bipedalism, 38–40, 97, 114–15, 127

Bohr, Niels, 102–5, 220n12
brain, bilateral structures of, 29, 95–96
Brandt, Line, 165–66
Broca's area, 47, 77, 121
Burke, Kenneth, 182

Calvin, William, 41, 43, 222n13
Carroll, Joseph, 17–18
Catholic Mass, 178–79
Chomsky, Noam, 23, 149, 218n2, 221n3, 230
Christian, David, 8–9
cognitive poetics, 2, 15–17, 26–29, 92
Coleridge, Samuel Taylor, 21
Common Coding Theory, 33–34
complementarity, 47, 91, 94–95, 101–3, 105, 111. *See also* dyadic pattern
Corballis, Michael, 37, 116, 122

De Waal, Frans, 97, 109–10
Deacon, Terrence, 24, 123–26, 143–44, 158, 180, 222n7, 222n9
deixis, 151, 155, 160, 172. *See also* pointing
Dickinson, Emily, 5, 110, 215n4
diegesis, 134, 151, 170, 184, 196, 226n5

INDEX

Donald, Merlin, 10, 46, 58–69, 75, 189–90, 214, 218n2; episodic stage, 41, 59, 62–69, 166; mimetic stage, 23–24, 41, 59–60, 69, 125, 141; mythic stage, 10, 59–62; theoretic stage, 10, 60, 62. See also memory systems: intermediate term
drama, 21, 73, 178–79, 182, 184, 208, 211
dream, 54–55, 67, 95, 192, 194–95, 198–99
Dual Coding Theory. See Paivio, Allan
Dual-Process Theory, 29–34, 56, 67, 95, 189, 199
duality of patterning. See Hockett, Charles F.
dyadic pattern, 19–22, 48, 55–57, 67, 83, 213–14; audition/vision, 210–11; bimanual coordination, 71, 167; communication, 135, 139, 142, 157, 181, 186, 188; figure–ground, 52, 93–94, 100, 130, 157, 160; metaphor and metonymy, 159–60; parallel–serial, 34–43, 47, 100, 102; perception–action, 33–34, 46, 49, 88; play, 73, 75, 79; semiotics, 114, 129

Egger, Victor, 74–75
egocentric. See spatial frames of reference
Eleusinian Mysteries, 178–80
Eliot, T. S., 4–5
emblem, 128–29, 136, 139, 172
Emerson, Ralph Waldo, 3, 171
emotion, 12, 16–18, 27, 92, 176, 191; aesthetic, 5, 73, 159, 197, 212; evolutionary adaptation, 51–53, 217n9; indexical sign, 110, 134, 170, 200
epic: genre, 35, 179, 182, 194, 196, 203, 211; Greek, 186–87, 196–97, 202, 207, 210, 212
episode, 37, 49, 51–52, 191–94, 197–98, 210; play, 63, 65–67, 73–74, 78
episodic memory. See Tulving, Endel
episodic stage. See Donald, Merlin
Evans, Jonathan, 30, 32

evocriticism. See Literary Darwinism
evolutionary psychology, 17–18, 216n12
extramission, 85
eye, structure of. See visual perception

fictive motion, 164–66, 173, 224n10
figure–ground perception. See dyadic pattern; visual perception
Fitch, William Tecumseh, 204–5
formulaic utterance, 117, 120, 137, 178, 202–5, 222n10; esoteric and exoteric language, 118–19, 143, 147, 183

Gallese, Vittorio, 47, 55, 163
genres, 181, 184, 207. See also drama; epic; ritual
gesture, 23–24, 66, 116–17, 125–26, 132, 134–35, 202–3; iconic, 73, 107, 119–22, 128–30; indexical, 113–16, 119–22, 127, 148, 180; symbolic, 136–39, 147. See also emblem
Gibson, James J., 92–93
Givón, Talmy, 119, 160–62
Goodale, Melvyn, 88, 90–93, 98, 133–34, 163–64, 220n9
Goody, Jack, 209
gossip, 35, 112, 137, 221n1. See also social information
Guiard, Yves, 39–40, 167

hand, 38–41, 127–28, 135, 146; eye–hand coordination, 39, 77, 92, 168; grips, 40–42, 121, 217nn4–5, 219n9; hand preference, 39–42, 167. See also gesture
hearing. See auditory perception
Heidegger, Martin, 70
hermeneutics, 11, 13–16, 27, 216n8
Hickok, Gregory, 134
Hockett, Charles F., 147, 149, 172, 223n1
hominid apes, 38–39, 59, 62, 86, 112–14
Homo: H. erectus, 96–97, 115–17, 128, 137–39; H. ergaster, 69, 114, 121, 147; H. habilis, 40, 109, 114, 121, 125, 147; H. heidelbergensis, 115–16, 119–20;

248

H. neanderthalensis, 111, 116, 119, 221n3; H. sapiens, 23, 41, 75, 111, 136–37, 147; H. sapiens sapiens, 111, 115–17, 119, 124–28, 181, 202, 213
Horace, 207–8
Hurley, Susan, 33–34, 42, 46
hypotaxis, 188, 209–10. See also parataxis

Iacoboni, Marco, 47, 121
image schemas, 52, 97, 153, 163–64, 168, 190–91
imagination, 22, 98–100, 110, 128, 163; kinematic, 46, 55; verbal, 168–70, 176, 188–89, 199, 211; visual, 54–55, 75, 87, 95, 153–55. See also image schemas; motor schemas
inscription. See writing
instinct, 48–49, 51, 106, 148
instrument, 7, 15–16, 21. See also tools
interference, 38, 199, 203, 210
invention. See rhetoric

Jakobson, Roman, 171–73, 188
James, William, 50, 102–3, 187, 219n3, 220n12
Jeannerod, Marc, 91, 219n6

Knight, Chris, 110–11, 119, 144
Koestler, Arthur, 217n2
Kosslyn, Stephen, 54, 86, 97–98
Kramnick, Jonathan, 18, 216n12
Krifka, Manfred, 167

Lakoff, George, 158, 162–63, 223n3, 224n8
Lancy, David, 63
Langacker, Ronald, 155, 157, 160, 166, 224n7
Leslie, Alan, 63–65, 79–80, 149, 218n3
lexicon, 20, 24, 53, 117–19, 147, 161, 172
Liberman, Alvin, 132
Lieberman, Philip, 221n3

linguistics, 11, 15, 27, 161–63, 181
Literary Darwinism, 17–19, 216n12, 224n2
lyric, 2, 176, 212. See also song

MacLeish, Archibald, 157
McCune-Nicolich, Lorraine, 149–50
McNeill, David, 135–36, 202
memory systems: episodic, 37, 45–46, 54, 61, 99–100, 161–62, 198; intermediate term, 35–37, 60–61, 73, 166, 185, 198; procedural, 54, 69, 76, 98; semantic, 53–54, 91–92, 97–100, 134, 161, 190–91; working, 36, 60, 63. See also rhetoric: memory
mental mapping, 44–45, 87, 97–98, 174, 189, 220n9. See also search behavior
mental walk, 174, 193. See also rhetoric: memory
Merleau-Ponty, Maurice, 71–72
metacommunication, 65–66, 74. See also Bateson, Gregory
metaphor, 67, 80, 151–52, 158–60, 162–63, 171–72
meter. See prosody
metonymy, 151, 171–73
Milner, David, 88, 90–93, 101, 134
mimesis, 61, 163, 170, 226n5. See also Donald, Merlin: mimetic stage
mind reading, 61, 65–66, 97, 126
mirror neuron system, 46–47, 73, 96, 162–64, 218n6
Mishkin, Mortimer, 90–93
Mithen, Steven, 76, 115, 137
motor schemas, 69, 164, 168, 191
motor theory of speech perception, 132–34
mouvance, 207
Müller, Friedrich Max, 22–24
multitasking, 37–47. See also parallel process
myth, 62, 185–86, 194–95. See also narrative; orality

INDEX

Napier, John Russell, 40–42
narrative, 35–37, 62–63, 112–14, 181–86, 192–98, 202–12
Neisser, Ulric, 130
Nietzsche, Friedrich, 198–99

object information, 11, 110–14, 117–20, 136–37, 142–43, 147. *See also* social information
Oldowan culture, 41, 147
Olson, Charles, 7
Ong, Walter, 185, 207
optic flow, 93, 162, 165, 199
oral-formulaic theory (Parry and Lord), 203–4
orality, 62, 133, 176, 180–212 passim, 214. *See also* epic; narrative; repetition

Paivio, Allan, 188, 225n5
paragesture, 139
paralanguage: gestural, 107, 135, 139–40, 202–3; vocal, 139–40, 200–201
parallel process, 30–32, 42–43, 47–52, 88–89, 100–103, 130–35, 166–67, 214. *See also* multitasking; serial process
parataxis, 188, 191, 206, 208–10. *See also* hypotaxis
parts of speech: modifiers, 154–55; nouns, 91, 112, 127, 137, 154–55, 168, 171; prepositions, 154; pronouns, 24, 78, 143, 145–46; verbs, 117, 127, 164
Peirce, Charles S., 124–25, 218n5. *See also* semiotics
personification, 170
Pinker, Steven, 23, 221n3
Plato, 2–5, 13, 52, 226n5
play, 21–22, 77–81, 218n4, 223n2; frames, 20–21, 25, 65–67, 74, 78, 197; imitative, 72–75; language, 148–52, 212; object, 63–67, 150–51; social, 20, 62–68, 72–74, 78
Poeppel, David, 134
poetics, 2, 10–17, 26–29, 216nn9–11

pointing, 113–15, 126–27, 146. *See also* deixis
Polanyi, Michael, 70–72
Pound, Ezra, 5, 159
pretense. *See* play
prosody, 26, 200–202
protolanguage, 107, 114–21, 129, 136–37, 185. *See also* formulaic utterance

repetition, 185–88, 197–98, 204, 206
rhetoric, 11–13, 15, 27, 142–43, 153, 168–69, 182, 215n7; arrangement, 168, 170, 182; delivery, 168, 174, 196–200, 203–5; invention, 168–69, 173–74, 197; memory, 168, 173–74, 189–96, 215n7; style, 168, 170–71, 183–85, 189, 208–10, 226n3
Rimbaud, Arthur, 4
ritual, 176–82, 196
Rizzolatti, Giacomo, 46–47
Russo, Joseph, 187

Schechner, Richard, 196
sclera, 97, 126
search behavior, 45, 49, 96–97, 156, 169, 217n6. *See also* mental mapping
semiotics, 20, 24, 115, 211, 122, 218n5; icon, 20, 66, 74, 78–79, 123–29, 136–40, 180, 203; index, 25, 66, 74, 113, 115, 123–29, 151–52, 171, 202; symbol, 19–21, 24–25, 124–26, 136–39, 144, 152, 211. *See also* gesture; lexicon; metaphor; metonymy; play; ritual
serial process, 30–32, 42–50, 96–103, 166–68, 185; sequential, 54–56; successive, 43–45, 53. *See also* parallel process
social information, 23, 110–14, 137, 142–46. *See also* object information
song, 177, 181, 212, 225n7
spatial frames of reference, 94–101, 162, 164, 173, 199, 220nn8–9
stone-knapping, 41, 69. *See also* tools

style. *See* rhetoric
syllogism, 210
synecdoche, 173
syntax, 24, 117–20, 137, 147

Talmy, Leonard, 156–60, 164–66, 224n9
tense, 136, 189, 192, 212
theory of mind, 60, 77–78, 97
Tomasello, Michael, 76, 121
tools, 21, 29, 39–43, 68–72, 98; found, 12, 21, 41, 68, 169, 193; made, 12, 27, 41, 68–72, 76–77, 169. *See also* instrument; verbal artifact
Tulving, Endel, 37, 45, 161, 190–91, 225n6

umwelt, 88–89, 94, 139
Ungerleider, Leslie, 90–93

verbal artifact, 22, 25–27, 181–85, 193–96, 213

visual paths. *See* visual streams
visual perception, 55–56, 82–105 passim, 122–23, 127, 153–54, 157, 164–65. *See also* imagination
visual streams: dorsal, 90, 93–96, 98–102, 161–65, 172–74, 199; ventral, 90–102, 161, 163–65, 173–74, 199
voice, 129–36, 139, 200–202, 210. *See also* auditory perception

Whitman, Walt, 4–7
Williams, William Carlos, 8, 157
Winnicott, Donald Woods, 64, 150
Wray, Alison, 117–19, 137, 143, 147, 203
writing, 13–14, 26, 173, 193, 206–14

Young, Richard, 39–41

Zatorre, Robert, 133
Zlatev, Jordan, 125–26
Zumthor, Paul, 204, 207

GPSR Authorized Representative: Easy Access System Europe, Mustamäe tee 50, 10621 Tallinn, Estonia, gpsr.requests@easproject.com